Interpreting Ecological Data

With over 70 exercises containing more than 270 questions, *Interpreting Ecological Data: Exercises for Undergraduates* allows students to apply ecological theory learned in classes to new, 'real' data from the scientific literature. Rather than simply rote learning word-perfect answers, students are empowered to apply their knowledge and develop critical thinking skills: core attributes for science undergraduates.

The book covers ecological principles such as predation, competition, mutualism, food webs and biogeochemical cycling. With numerous examples on each theme, students can gain plenty of practice across a wide range of topics, providing a solid foundation for their ecological knowledge and developing transferable skills in data interpretation. Almost all the exercises use 'real' data from published papers, sometimes messy and not always supporting the hypotheses they were designed to investigate—just like in real life! The answers are all included in the final chapter, with detailed explanations where relevant.

The exercises can also be used for assessment, enabling students to demonstrate their understanding, and lecturers can quickly convert them for use in lectures, workshops or assessments. Most importantly, *Interpreting Ecological Data: Exercises for Undergraduates* aims to stimulate student enthusiasm for ecology and what researchers actually do, starting an unexpected and exciting new journey.

Interpreting Ecological Data

Exercises for Undergraduates

Clare J. Trinder

CRC Press
Taylor & Francis Group
Boca Raton London New York

CRC Press is an imprint of the
Taylor & Francis Group, an **informa** business

Designed Cover Image: Shutterstock

Noisy miners (*Manorina melanocephala*) native to Australia; see exercise 4.8, page 64.

First edition published 2026
by CRC Press
2385 NW Executive Center Drive, Suite 320, Boca Raton FL 33431

and by CRC Press
4 Park Square, Milton Park, Abingdon, Oxon, OX14 4RN

CRC Press is an imprint of Taylor & Francis Group, LLC

ISBN: 978-1-041-08177-7 (hbk)
ISBN: 978-1-041-08174-6 (pbk)
ISBN: 978-1-003-64409-5 (ebk)

DOI: 10.1201/9781003644095

Typeset in Minion Pro
by Deanta Global Publishing Services, Chennai, India

*To the people who taught me how to teach,
especially Michelle Pinard, a friend and mentor,
who showed me how it was done.*

*To my many supportive colleagues, especially Cath
Dennis, Vasilis Louca and Sarah Woodin, who
frequently helped keep the show on the road.*

*And especially to the many students with whom I have
worked; your passion, dedication and commitment to
understanding ecology inspired me to find new and
challenging approaches to support your learning.*

Contents

About the author xi

Preface xiii
How to use this book xiv
Non-metric multidimensional scaling (NMDS) xx
Principal component analysis (PCA) xxii

Acknowledgements xxv

1 Biodiversity 1
Exercise 1.1: Biodiversity and stability 1
Exercise 1.2: Biodiversity and ecosystem function 4
Exercise 1.3: Biodiversity and invasibility 6
Exercise 1.4: Invasibility of plant communities 7
Exercise 1.5: Colonisation of islands 8
Exercise 1.6: Testing the Janzen–Connell hypothesis 9
Exercise 1.7: Herbivores and bird diversity 11
Exercise 1.8: Changes in species richness in chalk grasslands 14

2 Communities and disturbance 15
Exercise 2.1: Disturbance by an invasive freshwater fish 15
Exercise 2.2: Fynbos and fire 19
Exercise 2.3: Seed burial, fire, germination 21
Exercise 2.4: Disturbance to coral communities (1) 23
Exercise 2.5: Disturbance to coral communities (2) 25
Exercise 2.6: Salt marsh succession 26
Exercise 2.7: Rocky shore succession 28
Exercise 2.8: Invasive species and grassland succession 31

3 Energy, matter and decomposition 33
Exercise 3.1: Salamanders in streams 33
Exercise 3.2: Soils over time in Hawaii 37
Exercise 3.3: Coral spawning and nutrients 39

Exercise 3.4: Decomposition by woodlice 40
Exercise 3.5: Home-field advantage for decomposing litter? 42
Exercise 3.6: Plant litter and termites 43
Exercise 3.7: Whales and nutrient cycling 45
Exercise 3.8: Nutrients, mangroves and seabirds 48

4 Intraspecific competition and life histories 51
Exercise 4.1: Population cycles in Capitella capitata 51
Exercise 4.2: Intraspecific competition over time in box tree moths 53
Exercise 4.3: Limpet density 56
Exercise 4.4: Trade-offs in life histories 58
Exercise 4.5: Lack clutch size 61
Exercise 4.6: Limiting factors in wood pigeons 62
Exercise 4.7: Pecking order in oystercatchers 63
Exercise 4.8: Control of noisy miners 64

5 Interspecific competition 66
Exercise 5.1: Competition between pink muhly grass and native species in South Korea 66
Exercise 5.2: Interactions between bluebirds 67
Exercise 5.3: Competition between water fleas 69
Exercise 5.4: Interactions between forest ants 71
Exercise 5.5: Coexistence in seabirds 72
Exercise 5.6: Seasonal changes in coexistence in ducks 73
Exercise 5.7: Gause, Paramecia and Lotka–Volterra 74
Exercise 5.8: Interactions between mink and otters 75

6 Mutualism and parasitism 78
Exercise 6.1: Elephants, ants and scale insects 78
Exercise 6.2: Grazing by marsh periwinkles 81
Exercise 6.3: Marsh periwinkles, grazing and parasites 82
Exercise 6.4: Anemone symbiosis 85
Exercise 6.5: Invasive crayfish in North America 87
Exercise 6.6: Yucca and moths 88
Exercise 6.7: Impact of a parasitic plant on salt marshes 90
Exercise 6.8: Avian vampire flies and their impact on Galapagos finches 91

7 Predation and herbivory 95
Exercise 7.1: Brown bear diet in Sweden 95
Exercise 7.2: Population cycles of Arctic ground squirrels 97
Exercise 7.3: Invasive mongooses in Japan 98

Exercise 7.4: Skate and mussels in Strangford Lough 99
Exercise 7.5: Impact of invasive pythons 101
Exercise 7.6: Predator–prey cycles 103
Exercise 7.7: Oak masting and weevil predation 105
Exercise 7.8: Impacts of predation on a tadpole community 106

8 Food webs and complex interactions 108
Exercise 8.1: Goldenrod, grasshoppers and spiders 108
Exercise 8.2: Grazing effects of red and roe deer 111
Exercise 8.3: Lizards, homopterans, midges and shrubs 114
Exercise 8.4: Kruger food web 115
Exercise 8.5: Coastal food web 116
Exercise 8.6: Food webs and decomposition 122
Exercise 8.7: Metabarcoding to unscramble water birds' diets 124
Exercise 8.8: Response of soil food webs to defoliation 127

9 Impacts of climate change 130
Exercise 9.1: Birds on mountains 130
Exercise 9.2: Parasitic trematodes' response to increased temperature 132
Exercise 9.3: Mycorrhizas and interactions 134
Exercise 9.4: Damselfish interactions 136
Exercise 9.5: Eider in the Arctic 139
Exercise 9.6: Warming and drying in tropical forest soils 141
Exercise 9.7: Arctic food webs 143
Exercise 9.8: Impact of warming environments on limpets 146

10 References, answers and explanations 149
Preface 149
Chapter 1 150
Chapter 2 153
Chapter 3 159
Chapter 4 164
Chapter 5 169
Chapter 6 173
Chapter 7 179
Chapter 8 183
Chapter 9 190

About the author

Clare J. Trinder is Emeritus Senior Lecturer in the School of Biological Sciences, University of Aberdeen, UK. She worked as a research plant ecologist from 2007 to 2011 at the University of Aberdeen in collaboration with the James Hutton Institute, and has practical experience in both national and international nature conservation as an environmental education officer and conservation officer.

Preface

Some years ago, after another slow and often painful round of exam essay grading for an introductory ecology course, a colleague and I were comparing notes. We were both dismayed by the number of word-perfect answers quoting back to us the details of our lecture slides, but when asked to apply that knowledge to a new example, students were often left completely floundering. Why did they find it so hard to use the theories and hypotheses they had learnt when applied to a different situation, we wondered?

I continued to ponder this question. Applying theory to new settings was evidently difficult. But we simply assumed that they would acquire this ability as they studied; why weren't we teaching students how to do this? In fact, we weren't even giving them practice in interpreting data and applying knowledge until we got to the exam or other assessed coursework. We might ask questions in lectures to encourage students to think in the right direction, but while a few keen students might be willing to answer, most of the class avoided eye contact and waited for others to risk getting something wrong in front of the class.

The following year, I restructured the ecology course, using one of our three lecture slots each week as a data interpretation class. Students were given two questions, using examples that they had not previously seen but could be answered by applying ideas covered in that week's classes. Students were told to write two paragraphs that demonstrated their understanding of the core concept and how it related to the new example. At the end of the class, we talked through the answers, and students could assess how well they had done. To encourage attendance at and engagement with these classes, the end-of-semester exam comprised 20 'new' questions of which the students had to answer 12. I soon realised I was going to need a lot of material to provide enough questions for this approach: ten topics, two questions for class sessions, a pool of 40 questions for exams plus extras for revision practice. I quickly ran out of examples from textbooks and started scouring journals for appropriate material.

Later, I changed the format to use a number of multiple-choice questions (MCQs) followed by some short-answer questions (<100 words). This allowed instant online marking for MCQs and considerably faster grading of written answers. I also preferred the flexibility when grading written answers to be able to give partial marks when it was clear a student was on the right track but hadn't quite made all the connections.

Initially, we used closed-book exams. Later, we switched to using the same questions in open-book exams, allowing students to take in their own notes, books or use the internet. Questions were 'camouflaged' by changing the names of species and/or identifying areas and not disclosing the source of the material. There was no difference in average exam grades when we used closed- or open-book exams.

Students found the approach challenging, but many commented later in their studies that they were grateful we had introduced this style of question early on and could see the value of developing these skills as they progressed through their degrees.

This book includes many of the original questions I devised plus new ones using recent papers. Almost all the exercises use 'real' data from published papers. Consequently, the data are sometimes messy and don't always support the ecological theories and hypotheses they were designed to investigate. The answers are all included in the final chapter, and except for the more straightforward questions that only require reading details from graphs, there is an explanation of the answer. Generally, I have used the conclusions or points discussed in the original paper when writing the questions and answers; you can go back to the source material to check, change or adjust my questions and answers as you see fit. As a student, I hope the questions will stimulate your enthusiasm for ecology and how researchers go about uncovering details and increasing our understanding of the interactions between individuals and species. Where particular questions pique your interest, why not go to the original paper and find out more? It could be the start of an unexpected and exciting new journey!

How to use this book

This book is about interpreting graphs, tables, data and diagrams; using the information provided to make sense of data and applying ideas, theories, hypotheses and information that you are (presumably) learning about in introductory ecology courses. It's not about analysing data or making sense of the output from analytical tests—there are many excellent books that cover these aspects already. You may never use some of the analytical approaches described in this book, but it is still valuable to be able to interpret the presented results and make sense of what is shown.

This book can be used by students, working independently, who want to practise data interpretation and apply their ecological knowledge to new examples and situations. It can also be used by academics to structure lectures or tutorials. The chapters mostly stand alone, so you can access questions at any point rather than needing to work through them in order. However, the more complex topics appear later in the book, as some of these can involve several interactions. Interactions rarely occur in a vacuum: e.g. predators, parasites or mutualists can change competitive effects. I have allocated the examples I've used to the topics based on the key interaction involved but it may not be the only one occurring. For example, many of the exercises include effects of invasive species and these could have been grouped together in a single chapter, but I have included them in the chapters covering the key interactions involved. Don't forget to stand back from the questions as well and think about how species and interactions link more broadly across the topics.

Some of the multiple-choice questions require demonstration of background ecological knowledge. Others simply involve reading some data in a figure and seem very straightforward; in these cases, the aim is to check that the graph has been interpreted correctly and to provide some hints to get students thinking in the right direction for answering the following questions.

For the short-answer questions, I have indicated the number of words that will give an appropriate answer. This is to avoid students spending time writing very detailed answers when I had only expected a three-word response. There is, of course, plenty of scope to expand written answers to cover more ground in the same way that I originally used them. Some MCQs ask the same question but in slightly different ways, to allow different combinations of questions to be used. Some of the longer written answers are designed to encourage students to pull together the answers to the previous questions to give an overview of the topic.

I've mostly used English names for species but have provided scientific names for reference and have indicated if a species name has changed since the publication of the paper from which the data come.

For easier reference, I've included here some short overviews of the approaches and analyses used when displaying data and how to interpret multivariate analyses. You might be familiar with some or all of these, in which case, move on to the first exercises. If you are less sure, here are a few tips to guide you.

How to interpret error bars on graphs

You will likely already be learning about descriptive statistics and how to present different forms of data. Often a graph will use the mean or average of a set of replicates to compare something that has been measured. It is helpful to include a measure to show the spread of data around a mean; this allows us to see how well the data are represented by this mean without having to scrutinise the raw data. There are a few ways that this can be done: standard deviation (SD), standard error (SE) and confidence interval (CI).

You may have come across the rule of thumb often used by biologists, that if error bars for different factors overlap, there is unlikely to be any statistically significant difference between these factors. That will generally be true but the converse (also widely used), that if the error bars don't overlap there *is* likely to be a significant difference between factors, will not always be the case. Analytical models can take into account more aspects of experimental design than a simple graph generally can, and this affects results. It also depends on which type of error bar has been used, so you need to note this when looking at graphs; published figures *should* always give this information in the legend.

Table P.1 gives a summary of what can and can't be inferred from different error bars.

Table P.1 Rules of thumb for interpreting error bars

	Overlap	No overlap
Mean ± 1 SE	Non-significant	Nothing can be inferred
Mean ± 1 SD	Nothing can be inferred	Nothing can be inferred
95% CI	Nothing can be inferred	Significant

Source: Schneider, 2020.

Figure P.1 shows a subset of data from a litter decomposition experiment. Mesh bags containing dried leaves from heather (*Calluna vulgaris*) and common cottongrass (*Eriophorum angustifolium*) were buried in a peat bog at a high or low water table.

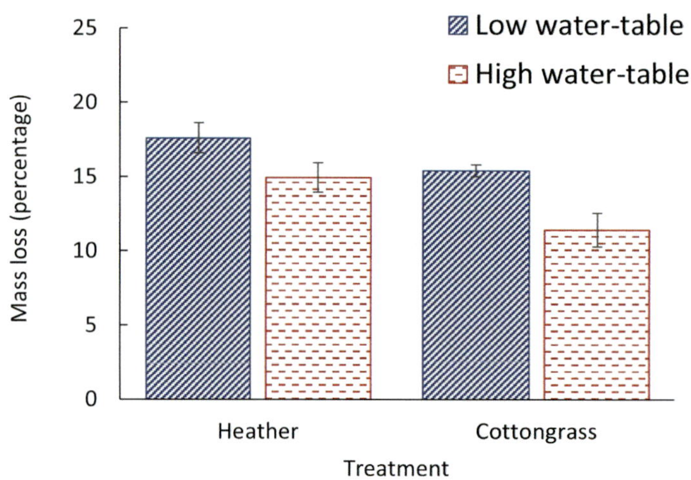

Figure P.1 Mass loss of heather and cottongrass litter at high and low water tables. Bars show means ±1 SE.

Figure P.1 uses standard errors; these show the smallest type of error bar. Using the rule of thumb of non-overlapping error bars, we might conclude that there is a difference between each species of litter buried at each water table.

If we look at the figure using confidence intervals (CIs), things look a little different (Figure P.2). These are 95% CIs, which means that 95 times out of 100, the true mean will fall between the upper and lower confidence intervals. The 95% CI is calculated from SE × 1.96. CIs can be calculated for a wide range of percentages, but the commonest is 95% and occasionally 99% (2.576 × SE). These error bars are bigger than the SEs.

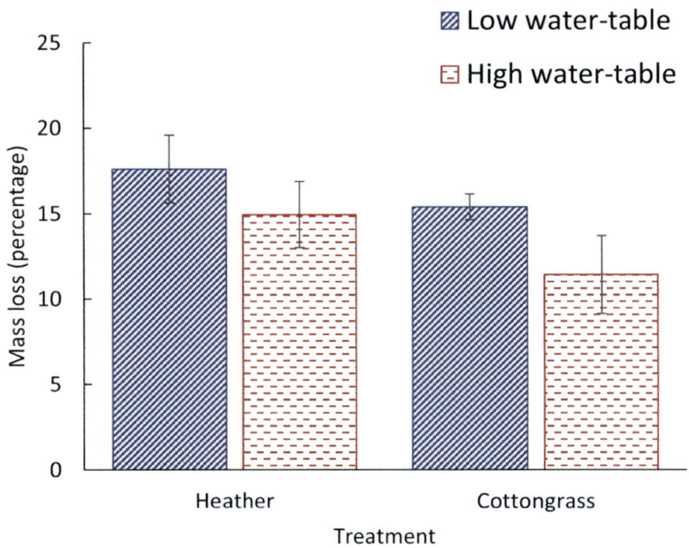

Figure P.2 Mass loss of heather and cottongrass litter at high and low water tables. Bars show means ±95% CI.

To get a clearer understanding of what is happening here, we need to do some statistical analysis. Table P.1 shows the results of a general linear model.

Table P.2 Results from a general linear model of mass loss data for two species at two water tables

Treatment	DF	Adj SS	Adj MS	F-Value	P-Value
Water table	1	48.972	48.972	9.41	0.006
Species	1	66.275	66.275	12.74	0.002
Water table × Species	1	2.641	2.641	0.51	0.484
Error	20	104.040	5.202		
Total	23	221.928			

Using the P-value column in Table P.2, we can see there are differences between mass lost at the two water tables and between the species, but the interaction between water table and species is also not significant (there's more detail on interactions later). The results confirm what we can also see from the graph—each species responds in the same way to the water table: more mass was lost at the low water table, less mass was lost at the high water table. The pairwise comparisons (Table P.3) give us more detail of what is happening.

Table P.3 Pairwise comparisons between species at both water tables; post hoc Tukey test with 95% confidence

	N	Mean	Grouping	
Heather, low water table	6	17.60	A	
Cottongrass, low water table	6	15.41	A	
Heather, high water table	6	14.94	A	B
Cottongrass, high water table	6	11.42		B

Note: Means that do not share a letter are significantly different from each other.

We can see which combinations were significantly different from each other, but how can we add these details to Figure P.2?

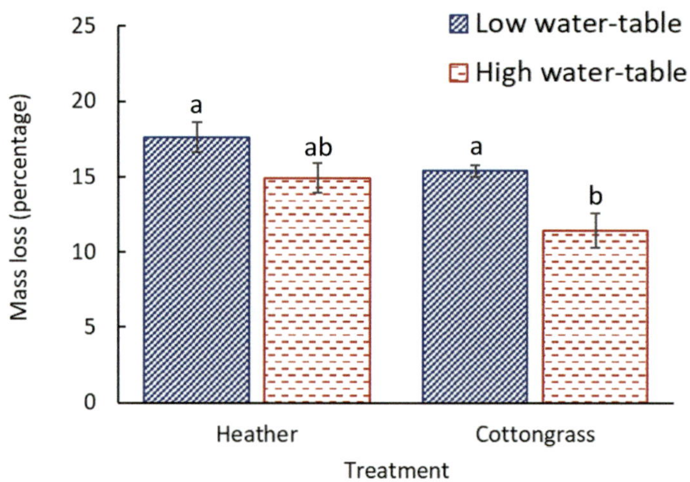

Figure P.3 Mass loss of heather and cottongrass litter at high and low water tables. Bars show means ±1 SE. Bars sharing a letter are not significantly different from each other.

From Table P.3 and Figure P.3 we can see that there was no difference in mass loss of heather between high and low water tables, nor between heather and cottongrass at the low water table (these all share the letter 'a'); there was also no difference in mass loss between cottongrass and heather at high water table (these bars share the letter 'b'), but there *was* a difference between the mass loss of cottongrass at the high water table and mass lost by heather and cottongrass at the low water table (these bars don't share a letter).

How should you interpret data where you are given the means and SEs but no other details? You can use the rule of thumb shown in Table P.1 in this situation but can only say that the figure 'suggests' a particular outcome.

Without the appropriate statistical test, you can't be confident about which, if any, factors are different from each other.

Many of the exercises in this book will use error bars of one type or another and some show results of statistical tests. Where there are no error bars or statistical results given, you should focus on the trends that are shown by the data.

The interaction in the preceding example was non-significant, but how might these data look if there were a significant interaction?

Table P.4 Results from a general linear model of mass loss data for two species at two water tables using adjusted data

Treatment	DF	Adj SS	Adj MS	F-Value	P-Value
Water table	1	131.63	131.628	29.40	<0.001
Species	1	106.71	106.709	23.83	<0.001
Water table × Species	1	283.78	283.784	63.38	<0.001
Error	20	89.55	4.477		
Total	23	611.67			

I've changed the raw data for cottongrass and there is now a significant difference between mass loss at the water tables *and* between the species, but the significant interaction shows that the species were not responding to the two water tables in the same way (Table P.4). In Figures P.1–P.3, we can see that cottongrass responded in the same way as heather to the water tables; it just lost less mass than heather. Figure P.4 shows that this was not the case with the amended data.

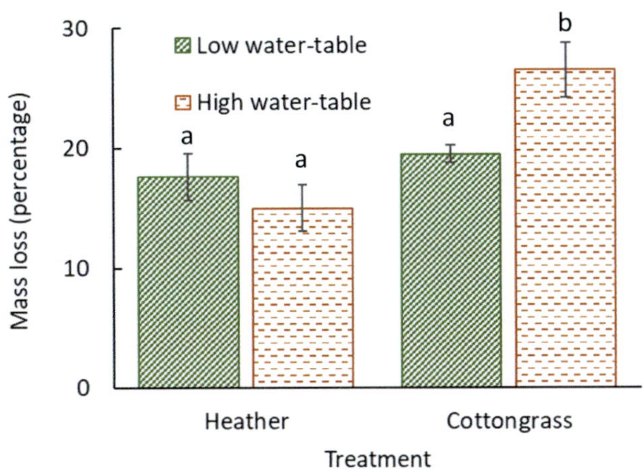

Figure P.4 Amended data showing a significant interaction of mass loss of heather and cottongrass at high and low water tables. Bars show means ±95% CI. Bars sharing a letter are not significantly different from each other.

While heather responded to high water table with a reduction in mass loss, cottongrass responded with *greater* mass loss. As before, the extra details come from the pairwise comparisons, shown in Table P.5.

Table P.5 Pairwise comparisons between species at both water tables using amended data; post hoc Tukey test with 95% confidence

	N	Mean	Grouping	
Cottongrass, high water table	6	26.50	A	
Heather, low water table	6	17.60		B
Cottongrass, low water table	6	15.41		B
Heather, high water table	6	14.94		B

Note: Means that do not share a letter are significantly different from each other.

Figure P.4 uses the results from Table P.5 (I've swapped the 'a' and 'b' around as the convention is to start the lettering from the left-hand side).

How to interpret results from multivariate analysis (MVA)

A few questions in the book use multivariate analyses such as non-metric multidimensional scaling (NMDS) or principal component analysis (PCA). They are part of a range of analytical approaches known collectively as multivariate analysis (MVA). These analytical techniques are complex, but even if you never carry one out, it is useful to be able to interpret the results. MVA simplifies complex (i.e. multidimensional) information and looks for similarities between sets of data and whether measured variables help to explain the patterns. They are often used, for example, in plant surveys where quadrats record a range of plant species at different densities and environmental data on soil type, C content, moisture, etc. The data could be displayed as a matrix or large data table, but MVA allows us to analyse the whole dataset simultaneously to look for patterns.

Non-metric multidimensional scaling (NMDS)

NMDS is one type of ordination. A biplot is a typical output, showing a scatter of points (Figure P.5). Points which are closer together are more similar and may represent species or communities. Arrows show fitted vectors which start at the origin (0,0). 'Confidence ellipses' may be included on the biplot; these show the confidence interval around the centre ('centroid') of each group and help to show the degree of overlap between factors; the size of the confidence interval used should be specified. Using our ecological knowledge, we can try to deduce the most likely factors that result in the clumping or spread of data points.

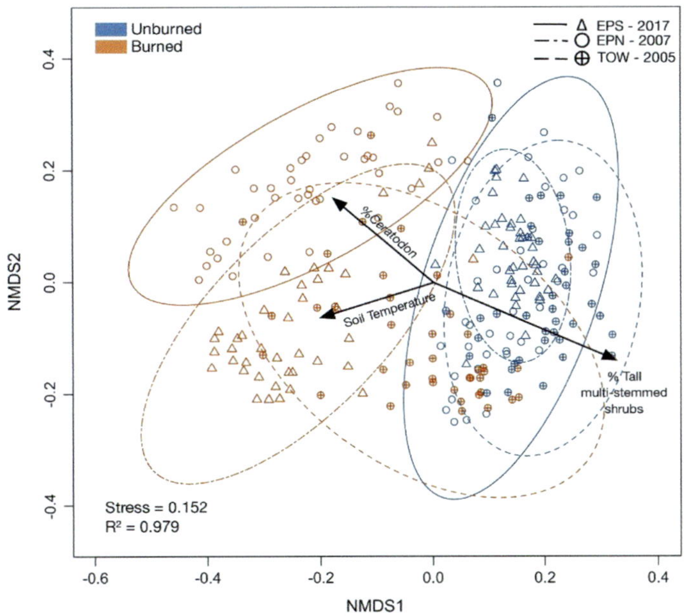

Figure P.5 NMDS biplot of spruce *Picea* spp. seedling emergence at treeline sites adjacent to tundra in subarctic Yukon. Points represent plant community functional groups at different sites, with different symbols to represent different sites surveyed at different times. Orange symbols are from burned sites and blue symbols are unburned. (From Brehault and Brown, 2022. Reproduced with permission, copyright 2022 Brehault and Brown, under exclusive licence to Springer Nature B.V.)

These analyses can be run with numerous axes ('dimensions,' given the letter 'k' and labelled in Figure P.5 as NMDS1 and NMDS2). Generally, the first two to three axes are the most useful when interpreting data (and most of us can't easily visualise patterns beyond three dimensions). The output from the analysis also gives a 'stress value,' indicating how well the analysis 'fits' the data: below 0.05 is considered excellent; anything over 0.2 is a poor fit. Adding extra dimensions to the analysis may decrease the stress value (i.e. improve the fit of the analysis to the data) but may not help you interpret the data. A high R^2 value (where included) confirms that the model is a good fit to the data. Note that not all papers provide the stress value. The numbers on the axes have no units and only signify differences between the points and groups.

What does Figure P.5 show? There is a stronger relationship between burned sites (orange points) and percentage cover of the moss, *Ceratodon purpureus* (the arrow runs through the confidence ellipses), as does the arrow for soil temperatures; unburned sites (blue points) have a stronger relationship with the percentage cover of tall, multi-stemmed shrubs (the arrow for multi-stemmed shrubs runs through the ellipse for the burned sites).

Figure P.6 shows another example with a slightly different representation of data. The outer points of each factor have been joined to create a polygon representing community composition of mites at different altitudes. We can see that there is some overlap between the community composition at 600 and 1600 m and also between 2000 and 2300 m but that the communities at 800 m and 2300 m are quite different. There is more spread along the x-axis than the y-axis (imagine how the polygons would look if they were collapsed down to one-dimension on either axis). Additional investigation by the researchers revealed that temperature and precipitation were the other main factors affecting the community composition of mites (Lin et al., 2023).

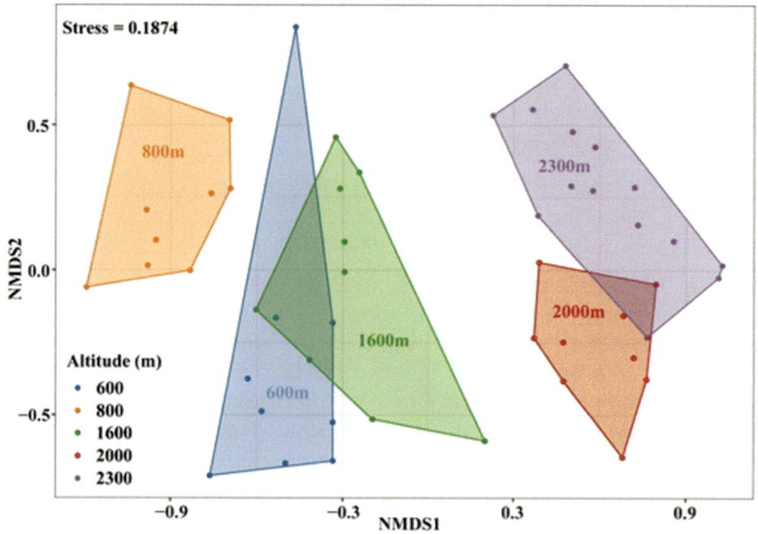

Figure P.6 Mite communities along an altitudinal gradient in Northeast China. Points represent the community composition of mites identified in individual samples. (From Lin et al., 2023. Reproduced under the Creative Commons CC BY licence. Copyright 2023 Lin et al., *Ecology and Evolution*, published by John Wiley & Sons Ltd.)

Principal component analysis (PCA)

PCA is similar to NMDS: a large, complex dataset is simplified to make it easier to see any broad patterns in the data. The analysis creates new, linear combinations of the data ('components' or 'dimensions') and these can be plotted on separate axes of a biplot. Each dimension is uncorrelated with the other components (but note that there are often correlations within the original data). There are as many components as there are variables in the dataset, but each subsequent variable explains less variation in the data than the previous one and generally the first two to three components are sufficient to show any ecologically meaningful patterns in the data. On a typical PCA biplot, points represent the sampling units (these could be communities, species assemblages or species). Units which are more similar to each other as calculated across the range of variables will appear closer together in the biplot. Explanatory variables are plotted as eigenvectors (arrows); the eigenvalues are the end point of each arrow. The percentage variation shown by each component is usually shown on the graph, either in the axis label or in the legend. As with NMDS, we need to use our ecological knowledge to make sense of what each axis represents (but be aware that sometimes the analysis can find patterns in data that don't have an ecological interpretation). Again, as with NMDS, the numbers on the axes have no units and only signify differences between the points and groups.

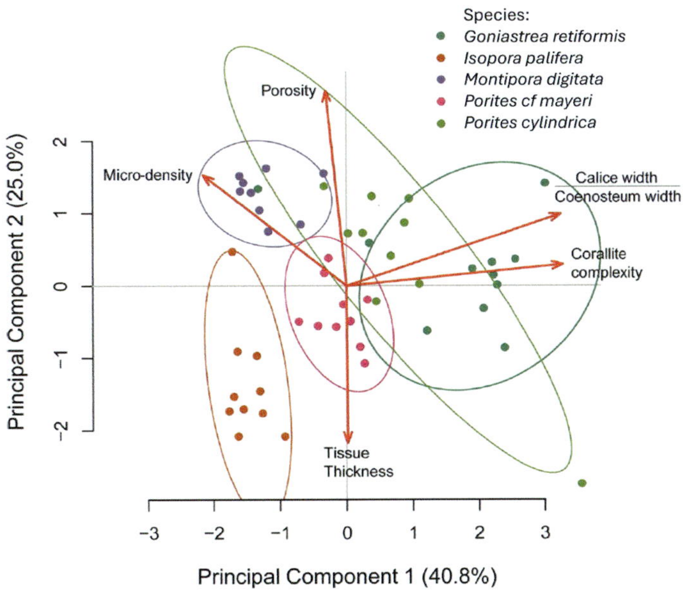

Figure P.7 PCA biplot of five coral species (colour coded) and their endolithic microbiome (microbial communities that live within the pore spaces of the coral structure). Points represent the microbiome samples recorded from each coral species. Five morphological features were also measured and these are shown as eigenvectors in red. (From Fordyce et al., 2021. Reproduced under the Creative Commons Attribution 4.0 International Licence. Copyright 2021 Fordyce et al.)

Axis 1 in Figure P.7 was positively influenced by corallite complexity and calice width/calice coenosteum and negatively influenced by microdensity; it explained 40.8% of the variation in the data. Axis 2 was positively influenced by porosity and negatively by tissue thickness and explained an additional 25% of the variation in the data. We can see that *Goniastrea retiformis* had a strong positive association with corallite complexity and calice width/calice coenosteum. In contrast, *Montipora digitata* had low corallite complexity but high microdensity.

Greenacre et al. (2022) provide a helpful overview of PCA, including a useful glossary of terms.

Acknowledgements

Many of the figures were digitised from the original papers using WebPlotDigitizer Version 4.8, distributed under the GNU Affero General Public Licence Version 3, copyright 2010–2024 Ankit Rohatgi.

1

Biodiversity

Exercise 1.1: Biodiversity and stability

There has been considerable debate for some decades over the role of biodiversity in 'stabilising' ecosystems (e.g. the system is able to maintain its species composition and functioning despite disturbance) and conferring 'resilience' (e.g. reducing time taken to recover following disturbance). One suggested mechanism for this is compensatory effects, such that if a disturbance reduces the abundance of one species, another in the community fills the vacated niche. Bai et al. (2004) investigated whether this occurred on grasslands in Inner Mongolia using data from 1980 to 2003. They identified vegetation types ('plant functional groups') based on the characteristics of the dominant species: perennial rhizome grass; perennial bunchgrasses; perennial forbs (flowering plants excluding grasses); shrubs and semi-shrubs; and annuals and biennials (Figure 1.1a and Table 1.1).

 DOI: 10.1201/9781003644095-1

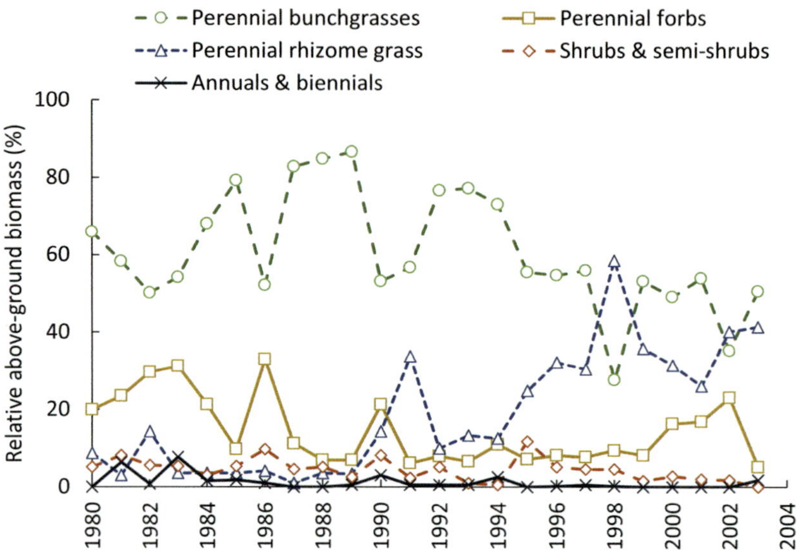

Figure 1.1a Relative above-ground biomass for five plant functional groups from 1980 to 2003. (Adapted from Bai et al., 2004. Reproduced with permission, copyright 2004 Macmillan Magazines Ltd.)

Table 1.1 Correlation coefficients between plant functional groups in relation to above-ground biomass

Plant functional groups	Correlation coefficient (***r***)	***P***
Perennial rhizome grass vs. perennial bunchgrasses	−0.4843	<0.0001
Perennial rhizome grass vs. perennial forbs	−0.3312	0.0002
Perennial rhizome grass vs. shrubs and semi-shrubs	−0.2989	0.0010
Perennial rhizome grass vs. annuals and biennials	−0.3189	0.0004
Perennial bunchgrasses vs. perennial forbs	0.0747	0.4197
Perennial bunchgrasses vs. shrubs and semi-shrubs	0.2516	0.0058
Perennial bunchgrasses vs. annuals and biennials	0.3001	0.0009
Perennial forbs vs. shrubs and semi-shrubs	0.0770	0.4050
Perennial forbs vs. annuals and biennials	0.2812	0.0020
Shrubs and semi-shrubs vs. annuals and biennials	0.1587	0.0848

Source: Adapted from Bai et al., 2004. Reproduced with permission, copyright 2004 Macmillan Magazines Ltd.

Question 1.1.1: Which pairs of functional groups show complementarity? (There are four correct answers.)

A. Perennial rhizome grass vs. Perennial bunchgrasses
B. Perennial rhizome grass vs. Perennial forbs
C. Perennial rhizome grass vs. Shrubs and semi-shrubs
D. Perennial rhizome grass vs. Annuals and biennials
E. Perennial bunchgrasses vs. Perennial forbs
F. Perennial bunchgrasses vs. Shrubs and semi-shrubs
G. Perennial bunchgrasses vs. Annuals and biennials
H. Perennial forbs vs. Shrubs and semi-shrubs
I. Perennial forbs vs. Annuals and biennials
J. Shrubs and semi-shrubs vs. Annuals and biennials

To assess stability, the researchers calculated the coefficient of variation (CV) in above-ground biomass:

$$CV = \left(Standard\ deviation/Mean\right) \times 100$$

It is used to compare variation in populations which cannot be compared directly due to differences in scale (Figure 1.1b).

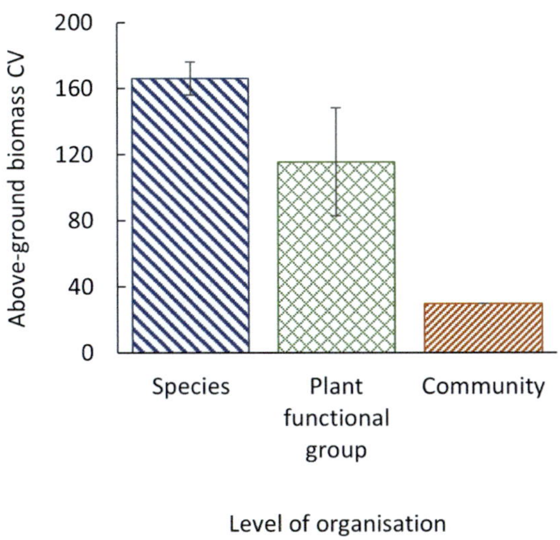

Figure 1.1b Above-ground biomass CV, calculated at the species level, using biomass from 16 commonly found species and compared against CV for plant functional groups and for the whole community. Bars show mean ± 1 SE. (Adapted from Bai et al., 2004. Reproduced with permission, copyright 2004 Macmillan Magazines Ltd.)

Question 1.1.2: Which level of organisation shows the highest stability?

A. Species
B. Plant functional group
C. Community

Exercise 1.2: Biodiversity and ecosystem function

Naeem et al. (1994) researched the effect of varying species richness at different trophic levels in a set of controlled environment 'ecotrons.' The experimental setup is shown in Figure 1.2a.

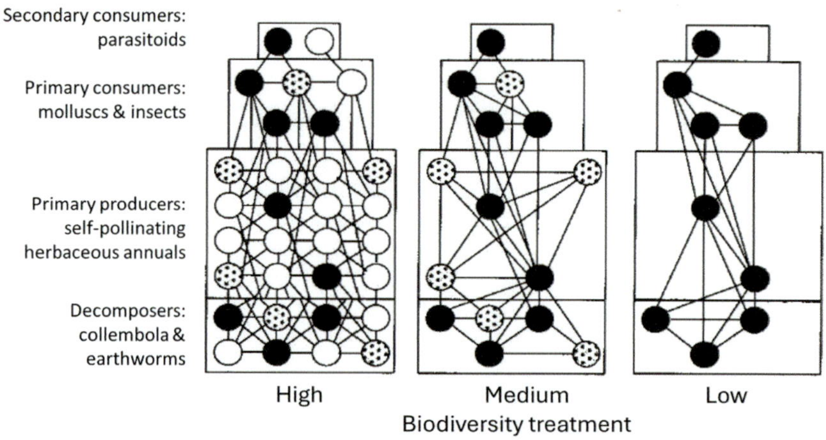

Figure 1.2a Schematic diagram showing the groups present in each treatment; note that the diagram is illustrative and doesn't represent physical locations or numbers of each group of species. Black circles represent species found in all biodiversity treatments; stippled circles are those found only in medium and high treatments; white circles are species included only in the high treatment; lines represent interactions between species. (From Naeem et al., 1994. Reproduced with permission, copyright 1994 Springer Nature Limited.)

The researchers measured a range of factors for each treatment, two of these are shown in Figure 1.2b.

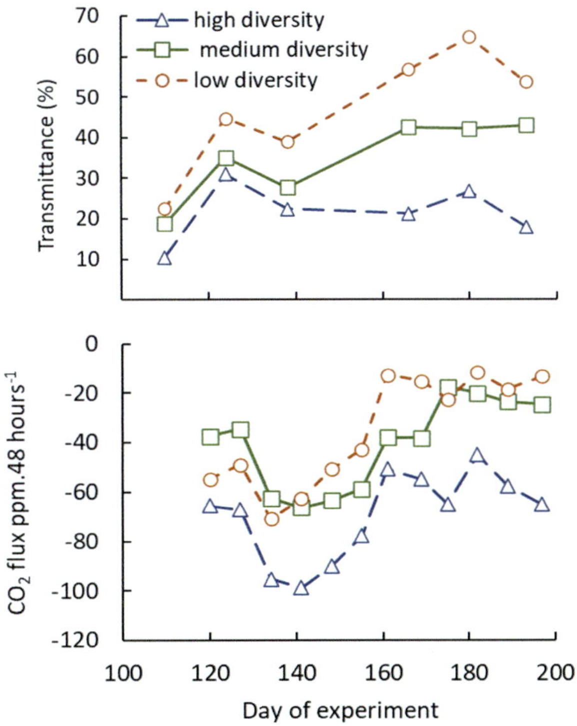

Figure 1.2b Top: Transmittance of photosynthetically active radiation (this is *inversely* correlated with productivity). Bottom: CO_2 flux (the difference between CO_2 taken in and given out; positive values show that more C is being respired than taken in). Both graphs show mean values. (Adapted from Naeem et al., 1994. Reproduced with permission, copyright 1994 Springer Nature Limited.)

Question 1.2.1: Which treatment has the highest productivity?

A. High diversity
B. Medium diversity
C. Low diversity

Question 1.2.2: CO_2 flux measurements in this experiment suggest that (two correct answers):

A. Even the most diverse system acted as a carbon source.
B. Only the most diverse system acted as a carbon sink.
C. The most diverse system accumulated the most carbon.
D. The most diverse system accumulated the least carbon.
E. High diversity is necessary to accumulate carbon within natural communities.
F. Even the least diverse treatment acted as a carbon sink.

Question 1.2.3: The high diversity treatment had a higher percent cover of vegetation and greater canopy architecture. How might this help to explain the results shown in Figure 1.2b? (No more than 40 words are needed.)

Exercise 1.3: Biodiversity and invasibility

Spotted knapweed (*Centaurea stoebe*; formerly known as *C. maculata*) is an invasive species of grasslands in the west of the US. Maron and Marler compared its ability to invade different mixtures of native grasses and flowers and monocultures (Figure 1.3).

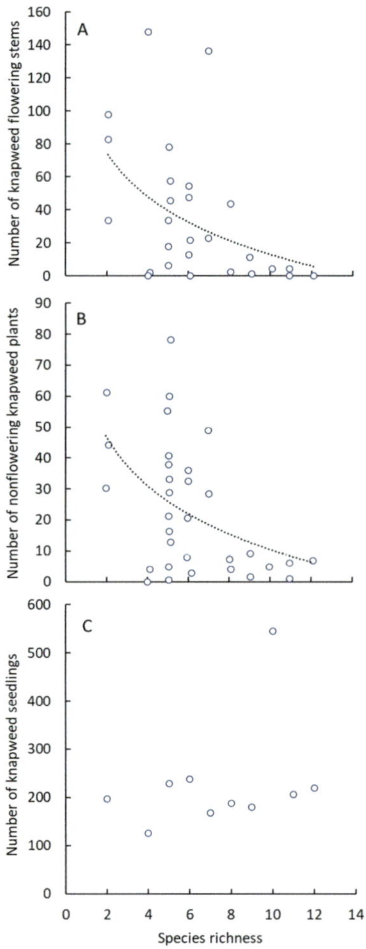

Figure 1.3 (A) Number of knapweed flowering stems. (B) Number of non-flowering knapweed plants. (C) Number of knapweed seedlings in mesocosms of different species richness (2–12 species). Curves show the line of best fit through the data through the data. (Adapted from Maron and Marler, 2007. Reproduced with permission, copyright 2007 by the Ecological Society of America.)

Question 1.3.1: Using the graphs, choose the most likely explanation of what is happening:

A. The more flowering knapweed stems within a community, the harder it is for other native species to persist.
B. The more non-flowering knapweed plants within a community, the harder it is for other native species to persist.
C. There is a negative correlation between numbers of knapweed plants and species richness.
D. Knapweed finds it harder to establish within more species-rich communities.

Question 1.3.2: Knapweed seeds germinate in autumn when many species in these communities are dormant. Explain the significance of this to the biodiversity/invasibility hypothesis. (No more than 60 words are needed.)

Exercise 1.4: Invasibility of plant communities

Researchers in New Zealand studied forests of native mountain beech (*Nothofagus cliffortioides*) over 23 years to look at patterns of invasion by a non-native species. Mountain beech forests naturally tend to have very low species richness. They recorded various factors including species richness of plots being invaded by non-native plants (Table 1.4).

Table 1.4 Percentage area of non-native plants in different species-richness classes

	Percentage area of non-native plants in plots	
Total plot species-richness class	**Mean**	**SE**
0–9	2.13	0.260
10–19	7.06	0.519
20–29	9.04	0.623
30–39	12.05	0.831
>40	16.10	2.026

Question 1.4.1: What does the diversity–invasibility hypothesis predict should occur in this type of community?

A. Diversity makes no difference to the ease of colonisation by invading species.
B. More diverse communities should be easier for invading species to colonise.
C. More diverse communities should be more difficult for invading species to colonise.

D. Communities of moderate diversity should be easier for invading species to colonise.
E. Every case should be assessed on an individual basis.

Question 1.4.2: The researchers noted that more diverse plots tended to have fewer dominant species. How might this affect the patterns they observed? (No more than 40 words are needed.)

Exercise 1.5: Colonisation of islands

Peter Becker looked at the patterns of diversity of beetles (Coleoptera) in the Canary Islands and Madeira. Figure 1.5 shows some of his results.

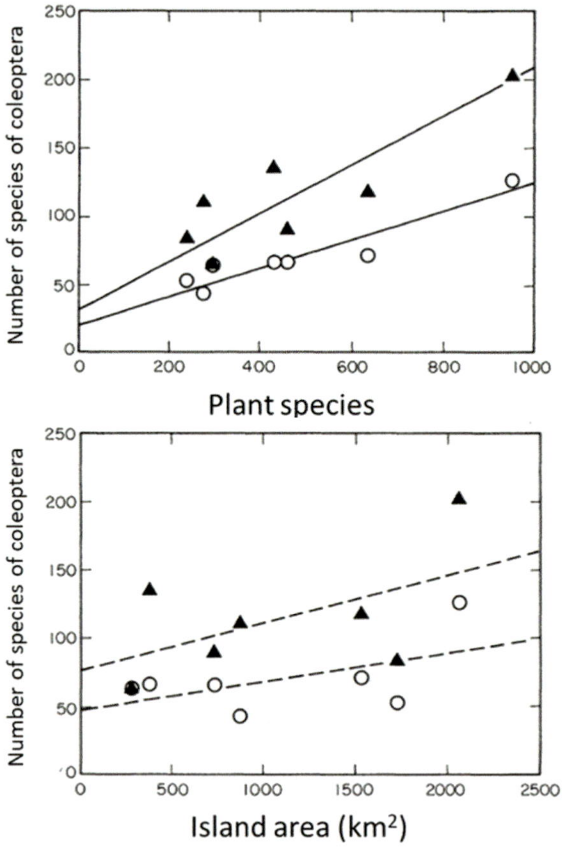

Figure 1.5 Diversity of beetle species in the Canary Islands and Madeira in relation to the number of plant species (top) and island area (bottom); triangles represent carnivorous species and circles represent herbivorous species. Regression analyses were significant for lines shown as continuous black lines and non-significant for the dashed lines.

Question 1.5.1: MacArthur and Wilson developed an equilibrium model to predict the numbers of species that would be found on islands. From their work, which of these options would you expect to have seen in relation to Becker's work on beetles?

A. A positive relationship between increasing numbers of plants and increasing beetle species
B. A positive relationship between number of plant species and size of island
C. A negative relationship between distance of island from the mainland and number of beetle species
D. A positive relationship between number of beetle species and area of island

Question 1.5.2: Which of these options would best explain the lack of fit with theory?

A. The number of plant species isn't dependent on the area of the islands.
B. Coleoptera are very small, so the size of island needs to be considered at a smaller scale than used here.
C. Herbivorous and carnivorous species tend to show different patterns of colonisation of new areas.
D. Beetles find it harder to colonise new areas, so species which go locally extinct are less likely to be replaced by new species arriving.

Figure used with permission of John Wiley & Sons - Journals, from Becker, P. 1992. Colonisation of islands by carnivorous and herbivorous heteroptera and coleoptera: effects of island area, plant species richness and 'extinction' rates. *Journal of Biogeography*, 19: 163–171. Permission conveyed through Copyright Clearance Center, Inc.

Exercise 1.6: Testing the Janzen–Connell hypothesis

The Janzen–Connell hypothesis proposes high diversity, particularly in tropical rainforests, occurs because seeds that land near their parent tree tend to be eaten by seed predators that specialise on that species. Seeds that are dispersed farther away are less likely to be encountered by specialist predators and are thus more likely to survive. Working in temperate forests in Indiana, Packer and Clay tested the Janzen–Connell hypothesis in black cherry (*Prunus serotina*) (Figure 1.6a).

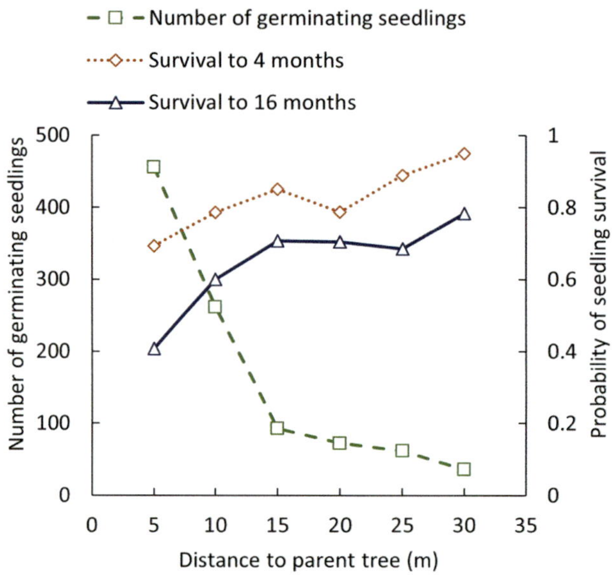

Figure 1.6a Relationship between distance from parent plant and number of germinating seedlings; probability of seedling survival to 4 months; and survival to 16 months. (From Packer and Clay, 2000. Reproduced with permission, copyright 2000 Macmillan Magazines Ltd.)

Question 1.6.1: Using Figure 1.6a, select the correct statement:

A. The number of seedlings germinating increases farther away from the parent tree, after both 4 and 16 months.
B. More seeds germinate close to the parent tree than farther away.
C. More seedlings germinate close to their parent tree because the soil conditions are likely to be more appropriate for them.
D. These data do not support the Janzen–Connell hypothesis.

Question 1.6.2: Approximately how many seedlings had survived after 16 months at a distance of 10 m from the parent plant?

A. 100
B. 150
C. 250
D. 300
E. 450

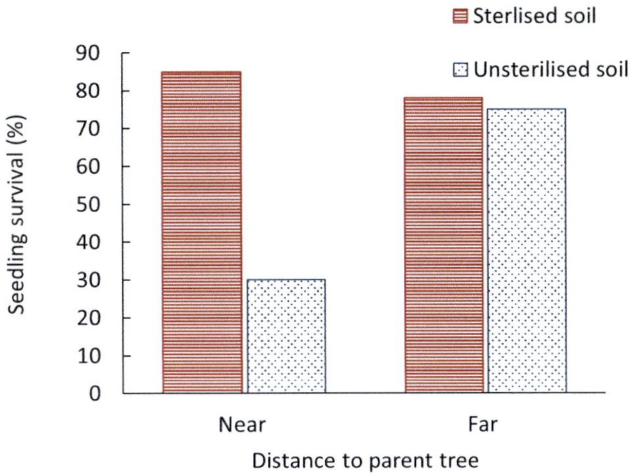

Figure 1.6b Effect of sterilising soil on seedling survival near to and far from parent trees. (From Packer and Clay, 2000. Reproduced with permission, copyright 2000 Macmillan Magazines Ltd.)

Question 1.6.3: Use the results shown in Figure 1.6b and those from the two previous questions in this exercise to select the most appropriate statement from the following options.

A. There is an advantage to growing close to a parent tree as the conditions are clearly suitable for its growth.
B. There is a clear effect of predation on seedlings which germinate too close to the parent tree.
C. There is a clear effect of invertebrate herbivory on seedling survival—those too close to their parent tree are easily found and eaten by invertebrate grazers.
D. Reduced survival close to parent trees is probably due to fungal soil pathogens that specialise on cherry.
E. Seedlings survive better farther from their parent tree due to reduced competition for the same resources.

Exercise 1.7: Herbivores and bird diversity

The Kenya Long-term Exclusion Experiment at the Mpala Research Centre, Laikipia, began in 1995. A series of fenced plots were established to allow comparison of grazing by different groups of animals. There were three treatments: grazing by megaherbivores and herbivores; just herbivores; or no grazing by wildlife. Megaherbivores in this system are reticulated giraffe and elephant. Herbivores include Grevy's and Burchell's zebra, Grant's gazelle, Jackson's hartebeest, eland, Besia oryx, steinbuck and Cape buffalo (Figure 1.7).

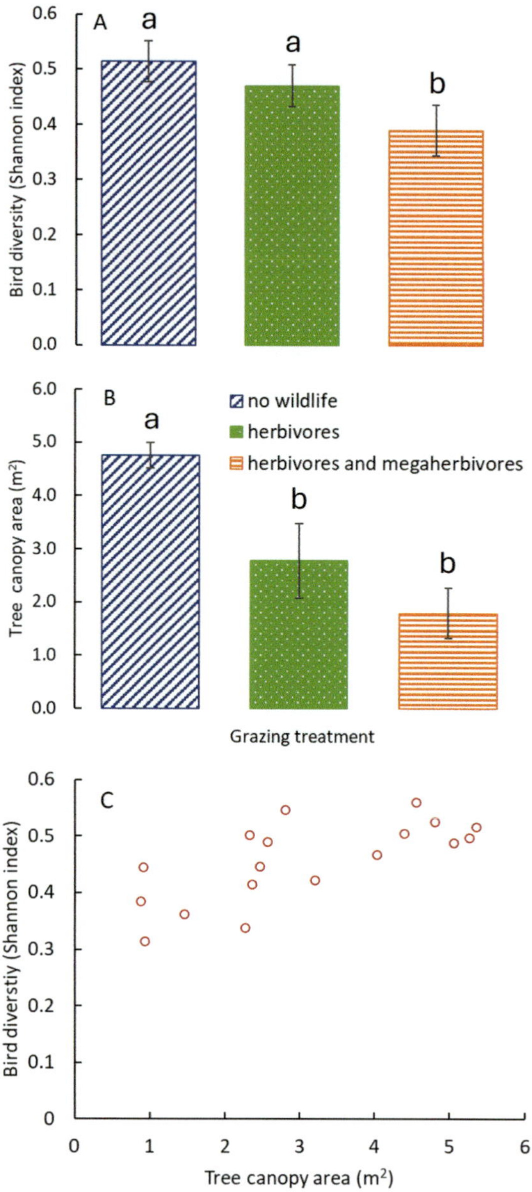

Figure 1.7 Effect of grazing treatments on bird diversity and tree canopy area. Graphs A and B show treatment mean ± 1 SE. Bars sharing a lowercase letter are not significantly different from each other. Graph C shows mean values. (From Ogada et al., 2008. Reproduced with permission, copyright 2008 Springer-Verlag.)

Question 1.7.1: What impact did megaherbivores together with herbivores have on bird diversity compared with no grazing wildlife and only herbivore grazing?

A. Increased diversity.
B. Significantly increased diversity.
C. Decreased diversity.
D. Significantly decreased diversity.
E. There was no difference.

Question 1.7.2: What impact did megaherbivores have on tree canopy compared with no grazing wildlife?

A. Increased diversity.
B. Significantly increased diversity.
C. Decreased diversity.
D. Significantly decreased diversity.
E. There was no difference.

Question 1.7.3: What impact did megaherbivores have on tree canopy compared with herbivore grazing?

A. Increased diversity.
B. Significantly increased diversity.
C. Decreased diversity.
D. Significantly decreased diversity.
E. There was no difference.

Question 1.7.4: Which of the following options best describes the relationship between bird diversity and tree canopy area?

A. Greater tree canopy area causes higher bird diversity.
B. Greater tree canopy area causes lower bird diversity.
C. Higher bird diversity creates greater tree canopy area.
D. Higher bird diversity creates lower tree canopy area.
E. There is a positive relationship between tree canopy area and bird diversity.
F. There is a significant positive relationship between tree canopy area and bird diversity.

Question 1.7.5: Explain the links between bird diversity, grazing by megaherbivores and tree canopy area. (No more than 50 words are needed.)

Exercise 1.8: Changes in species richness in chalk grasslands

Chalk grasslands are renowned for having very high species diversity. Figure 1.8 shows data from a chalk grassland in the Netherlands. This dataset shows the change in species richness over 15 years following cessation of grazing management in a single plot measuring 2 × 5 m. Heath false brome (*Brachypodium pinnatum*) is a native grass species that is usually found at low frequencies in chalk grassland.

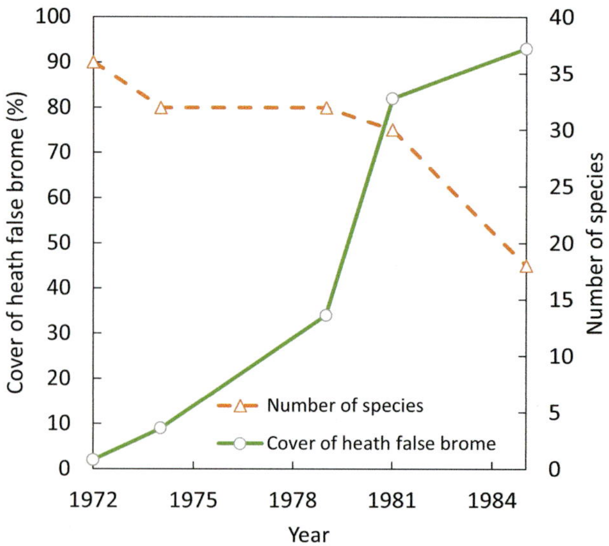

Figure 1.8 Changes in species diversity (right-hand y-axis) and percentage cover of heath false brome (left-hand y-axis) after grazing was abandoned on a chalk grassland site.

Question 1.8.1: Select the correct interpretations of the graph (more than one correct answer):

A. The number of species declined from approximately 90 in 1972 to fewer than 20 by 1985.
B. The number of species declined from over 35 in 1972 to fewer than 20 by 1985.
C. The cover of heath false brome increased from 1% to 2% in 1972 to over 90% by 1985.
D. The cover of heath false brome increased from 1% to 2% in 1972 to over 35% by 1985.

Question 1.8.2: Did the increase in heath false brome drive the decrease in species diversity or did the decrease in diversity allow false heath brome to increase? Explain your reasoning. (No more than 60 words are needed.)

2

Communities and disturbance

Exercise 2.1: Disturbance by an invasive freshwater fish

There are at least 17 non-native fish found in Everglades National Park. The African jewel fish (*Rubricatochromis letourneuxi*), a cichlid, established in Everglades National Park in 2000. It preys on small fish and invertebrates. For its first decade in the national park, jewel fish were found at low density, but between 2012 and 2017, the population density increased fivefold ('boom' phase). Populations then decreased in 2018 ('bust' phase). Researchers based at Florida International University used datasets from 1978 to 2021 to investigate changes in the resident fish populations and their recovery after jewel fish declined.

As part of the data analysis, the researchers used a technique known as non-metric multidimensional scaling (NMDS) (Figure 2.1a). See the Preface for background on what this technique does and how to interpret results generated by these analyses.

 DOI: 10.1201/9781003644095-2

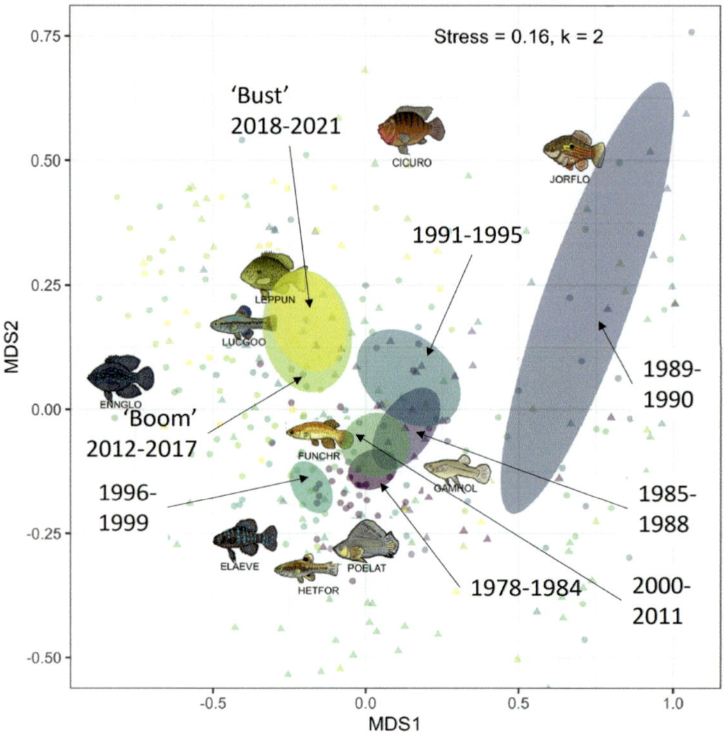

Figure 2.1a Fish community dynamics using NMDS. Each point represents a community assemblage determined by the relative abundances of each fish species; triangles and circles distinguish the data from two sites used for the analysis. The coloured ellipses are calculated around the mean x and y values using 95% confidence intervals for each time period. Community assemblages that are more similar to each other will be closer together. The ten commonest fish species are also shown on the figure and their positions show with which time period they are most commonly associated. (From Flood et al., 2024. Reproduced with permission, copyright 2024 Flood et al., under exclusive licence to Springer-Verlag GmbH Germany, part of Springer Nature.)

Fish names are mostly abbreviated from their scientific name, as shown in Table 2.1.

Table 2.1 Fish names and abbreviations

English name	Scientific name	Abbreviation
Bluefin killifish	*Lucania goodei*	LUCGOO
Blue-spotted sunfish	*Enneacanthus gloriosus*	ENNGLO
Eastern mosquitofish	*Gambusia holbrooki*	GAMHOL
Everglades pygmy sunfish	*Elassoma evergladei*	ELAEVE
Flagfish	*Jordanella floridae*	JORFLO
Golden topminnow	*Fundulus chrysotus*	FUNCHR
Mayan cichlid	*Mayaheros urophthalmus*	CICURO
Sailfin molly	*Poecilia latipinna*	POELAT
Spotted sunfish	*Lepomis punctatus*	LEPPUN
Least killifish	*Heterandria formosa*	HETFOR

Question 2.1.1: How many dimensions were specified when this analysis was run?

A. 1
B. 2
C. 5
D. 16
E. As many as needed until 100% of the variation was explained
F. As many as the software considered necessary

Question 2.1.2: The stress for this analysis suggests that:

A. The results are an excellent fit for the data.
B. The results are a good fit for the data.
C. The results are a poor fit for the data.
D. The results should be considered with great caution.
E. The fit is so poor that this paper should never have been accepted for publication.

Question 2.1.3: Which two species are most associated with the 'boom' and 'bust' phases of the jewel fish invasion?

A. Bluefin killifish
B. Blue-spotted sunfish
C. Eastern mosquitofish
D. Everglades pygmy sunfish
E. Flag fish
F. Golden topminnow
G. Least killifish
H. Mayan cichlid
I. Sailfin molly
J. Spotted sunfish

Question 2.1.4: There was a severe drought from 1989 to 1990. What effect did this have on the fish community? (No more than 30 words are needed.)

Question 2.1.5: Prior to the boom and bust of the jewel fish, this fish community composition was highly stable and showed very little change, with the exception of the 1989–1990 drought. True or false?

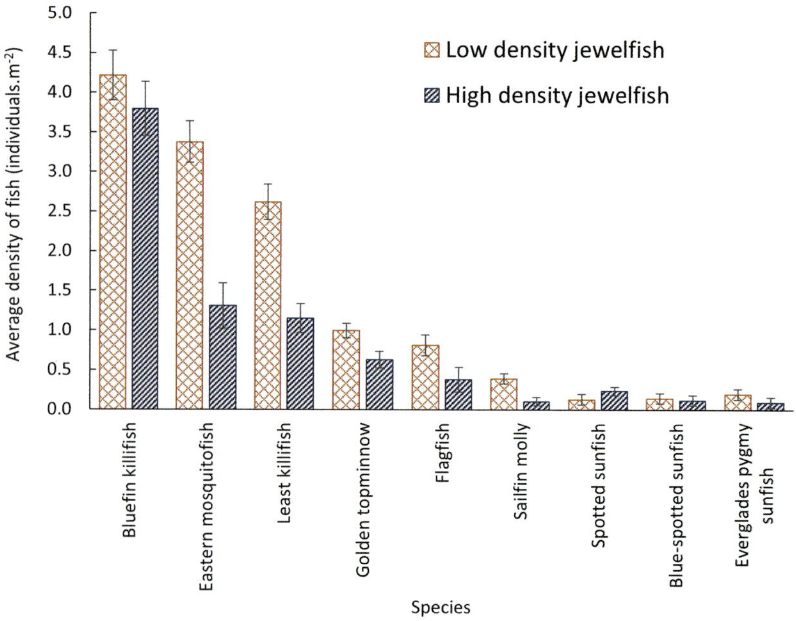

Figure 2.1b Mean fish density (± 2 SD) for each species during low-density (bust) and high-density (boom) phases. (Adapted from Flood et al., 2024. Reproduced with permission, copyright 2024 Flood et al., under exclusive licence to Springer-Verlag GmbH Germany, part of Springer Nature.)

Question 2.1.6: Use Figure 2.1b to determine the five fish species that drove the differences in community assemblage between the high and low density of jewel fish.

A. Bluefin killifish
B. Blue-spotted sunfish
C. Eastern mosquitofish
D. Everglades pygmy sunfish
E. Flag fish
F. Golden topminnow
G. Least killifish
H. Mayan cichlid
I. Sailfin molly
J. Spotted sunfish

Question 2.1.7: How much recovery was shown by the fish community in the 4 years after the boom phase of the jewel fish population? How could you explain this? (I've given three possible reasons, each explained in 30 words.)

Question 2.1.8: Blue-spotted and pygmy sunfish are smaller than jewel fish and eat omnivorous invertebrates. Spotted sunfish are larger than jewel fish and eat detritus, algae, and small invertebrates and fishes. How might these differences affect their interactions with jewel fish? (No more than 40 words are needed.)

Exercise 2.2: Fynbos and fire

Fynbos is a type of vegetation found only in the Cape region of South Africa. The vegetation comprises shrub and heath and is renowned for exceptionally high diversity and endemism. Fynbos is a fire-adapted community and is generally burned every 20 years to prevent succession. Proteas are a critical part of the shrub component of these communities; larger species of protea are unable to reach maturity if they are burned too frequently. Figure 2.2a shows results of an investigation of diversity of communities in relation to frequency of burning.

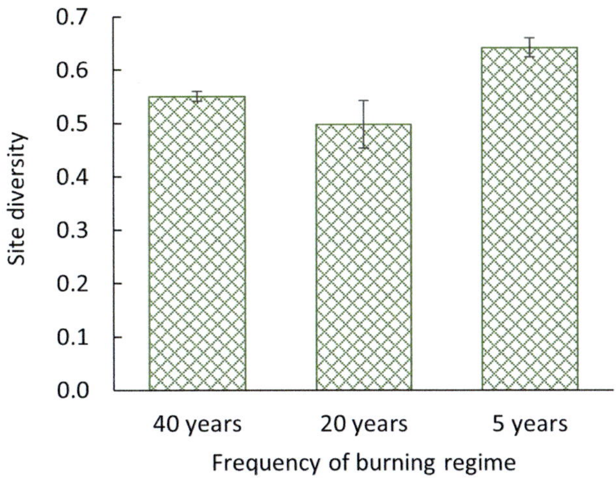

Figure 2.2a Site diversity in relation to burning regime, calculated at a 1 m[2] scale. Bars show mean ± 1 SE. (Adapted from Schwilk et al., 1997. Reproduced with permission, copyright 1997 Kluwer Academic Publishers.)

Question 2.2.1: Which of the following statements best describe the findings of the researchers? Select three.

A. Diversity is highest when burning happens rarely.
B. There was little difference between diversity on sites burned every 20 and 40 years.
C. Diversity was greatest at intermediate levels of burning.
D. The results show a good fit with the Intermediate Disturbance Hypothesis.

E. Diversity was greatest at very frequent levels of disturbance.
F. The results show the opposite effect expected from the Intermediate Disturbance Hypothesis.

The researchers also considered how diversity changed according to the scale at which it was measured. Figure 2.2b shows their findings.

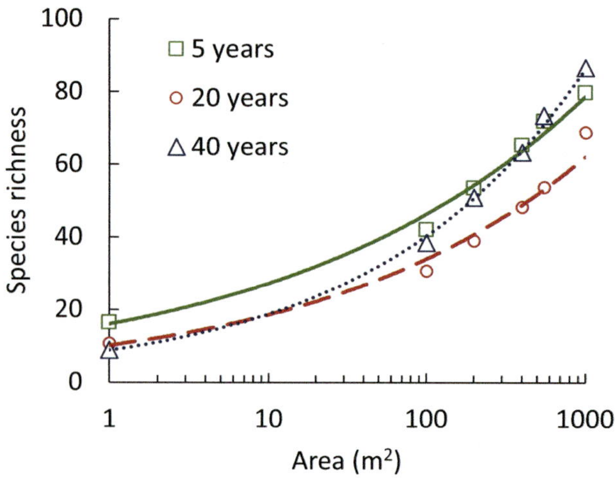

Figure 2.2b Mean species richness measured at different scales for three burning regimes. The curves are best fit lines from polynomial regression analysis. Note log scale for the x-axis. (Adapted from Schwilk et al., 1997. Reproduced with permission, copyright 1997 Kluwer Academic Publishers.)

Question 2.2.2: Use the figures to explain how diversity changes for the different burning regimes in relation to the scale of measurement. Do the data in Figure 2.2b fit with those in Figure 2.2a? (No more than 40 words are needed.)

Question 2.2.3: Why might very frequent fires result in higher diversity? (No more than 40 words are needed.)

Question 2.2.4: The protea has cultural importance in South Africa; for example, it is the logo of the national cricket team. What is the most appropriate management regime for conservationists to implement on fynbos sites?

A. Maximise diversity.
B. Maximise development of proteas.
C. Management priorities should be determined on a site-by-site basis.
D. Maximise diversity and development of proteas.

Exercise 2.3: Seed burial, fire, germination

A researcher in Australia investigated the relationship between the depth at which seeds of sweet wattle (*Acacia suaveolens*) were buried by ants and the impact of heat of subsequent fires on germination rates. Figure 2.3a shows some of his results.

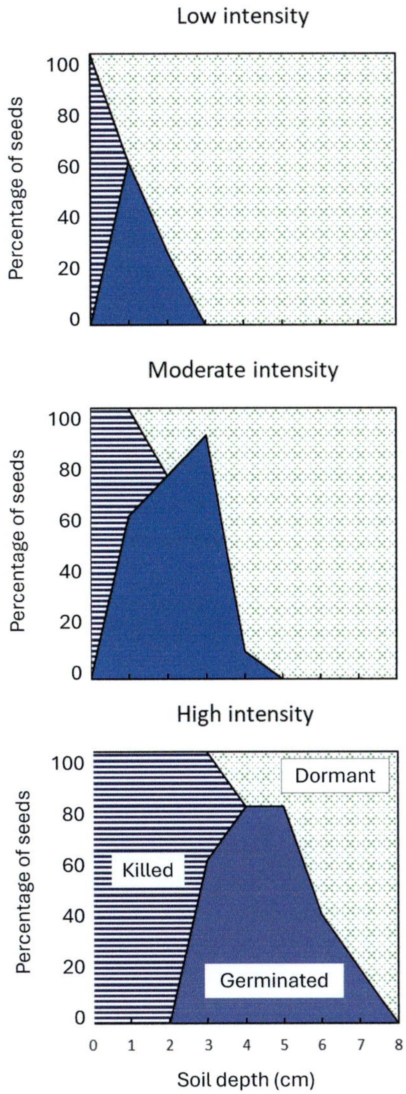

Figure 2.3a Percentage area stacked graphs showing effects of low-, moderate- and high-intensity fires on germination of sweet wattle at different soil depths. Explanation of colour coding on the bottom graph is the same for the low- and moderate-intensity fires. (Adapted from Auld, 1987. Reproduced with permission, copyright Ecological Society of Australia.)

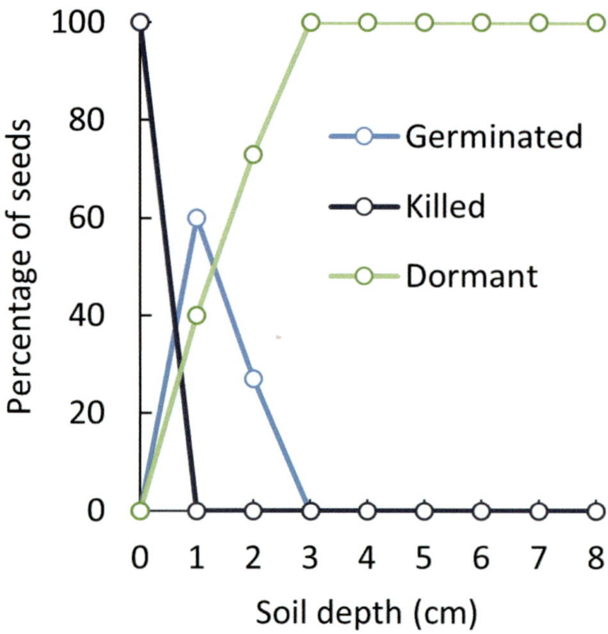

Figure 2.3b Line graph showing effect of low-intensity fires on germination of sweet wattle at different soil depths. (Redrawn from Auld, 1987. Reproduced with permission, copyright Ecological Society of Australia.)

Percentage area stacked graphs can be difficult to interpret. Compare Figure 2.3b with the corresponding graph in Figure 2.3a to check that you understand how these graphs work and what they are showing.

Question 2.3.1: Which of these statements relating to fire intensity are correct? (More than one answer is correct.)

A. All intensities of fire kill some seeds.
B. High-intensity fires killed the most seeds.
C. Moderate-intensity fires killed the most seeds.
D. In low-intensity fires, all seeds on the surface were killed.
E. In moderate-intensity fires, all seeds on the surface were killed.

Question 2.3.2: Which of these statements relating to seed germination is correct?

A. Fire is necessary for seeds of sweet wattle to germinate.
B. After high-intensity fires, a greater percentage of seeds germinate than after low and moderate intensity first.
C. High-intensity fires stimulate a greater *number* of seeds to germinate.
D. Moderate-intensity fires were best at stimulating germination.
E. Low-intensity fires were best at stimulating germination.

Question 2.3.3: Which of these statements relating to burial and seed survival is correct?

A. Low-intensity fires stimulated seeds to become dormant.
B. Seeds need to be buried at least 2 cm deep to be protected from all intensities of fire.
C. Only seeds buried at least 4 cm deep are protected from all intensities of fire.

Question 2.3.4: Frequent, low-intensity fires are sometimes favoured as a management tool. This reduces build-up of flammable material; in contrast, fire suppression can result in infrequent but very high-intensity fires. In this Australian plant community, what would happen to sweet wattle in the longer run if it were managed by frequent, low-intensity fires? What additional information would help to answer this question? (No more than 80 words are needed.)

Exercise 2.4: Disturbance to coral communities (1)

Space on reefs provides access to light and nutrients, so competition between corals for space can be intense. There are two ways in which corals can compete: directly, by digesting or growing on top of immediate neighbours; or by overtopping other corals without direct contact and thus reducing their access to light. A study on *Acropora* and *Goniastrea* species at Lizard Island on the Great Barrier Reef over 5 years assessed their competitive abilities. Species that were good at overtopping grew faster, were more vulnerable to being dislodged and produced more offspring (Figure 2.4).

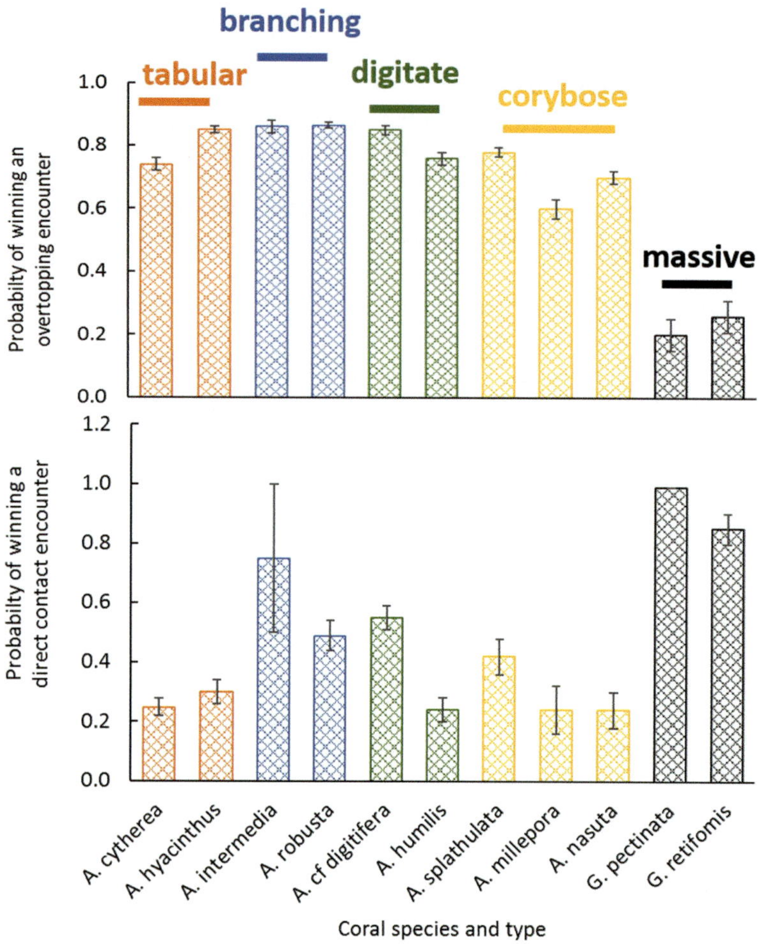

Figure 2.4 Competition between corals: probability of winning (top) an overtopping encounter and (bottom) a direct contact encounter. Bars show mean ± 1 SE; colour coding shows different growth forms of corals. (Adapted from Álvarez-Noriega et al., 2018. Reproduced with permission, copyright 2018 by the Ecological Society of America.)

Question 2.4.1: In the interactions between the coral species described above, there are two trade-offs. What are they?

Question 2.4.2: In no more than 40 words, how does disturbance from storms or wave action and these trade-offs affect coral reef diversity?

Exercise 2.5: Disturbance to coral communities (2)

The volcanic island of Moorea in the South Pacific is surrounded by a barrier reef. In 2008 some sections of the reef were affected by outbreaks of crown-of-thorns starfish (*Acanthaster planci*), a species that feeds on coral, especially branching and table types. The island was also hit by Cyclone Oli in 2010, which affected exposed parts of the reef system more than those in sheltered areas (Figure 2.5a).

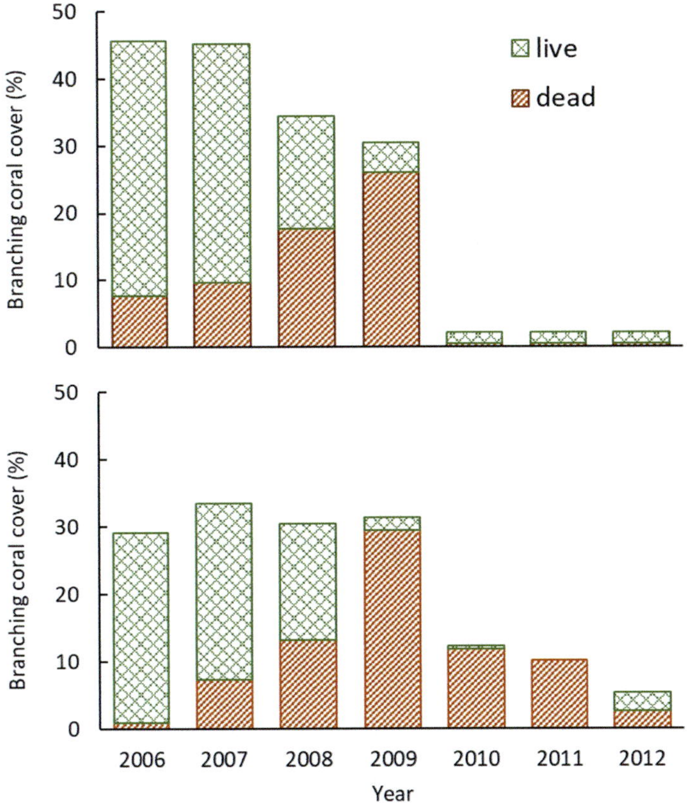

Figure 2.5a (Top) Proportions of live and dead corals from 2006 to 2012 on a reef affected by cyclone Oli and crown-of-thorns starfish. (Bottom) An area affected only by the crown-of-thorns starfish with minimal cyclone damage. (Adapted from Adam et al., 2014. Reproduced with permission, copyright 2014 Springer-Verlag Berlin Heidelberg.)

Question 2.5.1: Compare the graphs in Figure 2.5a showing the reef affected only by the crown-of-thorns starfish and that affected by the starfish and the cyclone. What is the main difference in relation to the structure of the reef between these areas? What implications might this have for reef recovery? (No more than 30 words are needed.)

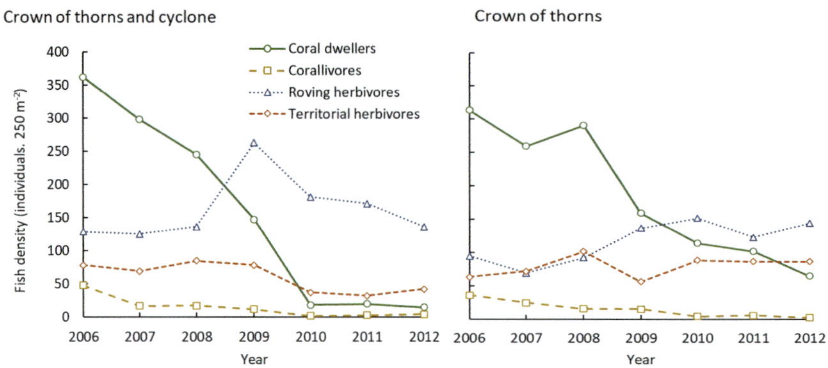

Figure 2.5b Change in abundance of four types of reef fish: coral dwellers, corallivores, roving herbivores and territorial herbivores between 2006 and 2012. Points show mean values. (Adapted from Adam et al., 2014. Reproduced with permission, copyright 2014 Springer-Verlag Berlin Heidelberg.)

Question 2.5.2: Why might roving herbivores have increased following the cyclone and in response to the outbreak of crown-of-thorns starfish? (More than one correct answer.)

A. This group of fish exerts a strong top-down effect on algae.
B. They were previously outcompeted by coral-dwelling fish.
C. They are less dependent on the structure of the coral.
D. There was more food available to them associated with the coral rubble.
E. They tend to breed and recruit in the more sheltered parts of the island (e.g. in lagoons).

Question 2.5.3: Roving herbivores are generally responsible for preventing algae from becoming dominant on coral reefs. What are the implications of this for the recovery of coral reefs on Moorea? (No more than 40 words are needed.)

Exercise 2.6: Salt marsh succession

Salt marshes are an important coastal ecosystem that experience regular tidal inundation; they are dominated by low-growing shrubby vegetation and grasses. They provide valuable protection to adjacent land by accumulating sediment and buffering against tidal surges and storms. High spring tides can deposit rafts of dead plant material such as seaweed, known as 'wrack,' to the higher sections of salt marshes, killing the plants beneath and creating bare patches. These areas are usually colonised by annual species or plants growing clonally, from adjacent vegetated patches. These plants ameliorate high salinity in the soils caused by evapotranspiration in the absence of plants.

In a salt marsh system in the lagoon of Mar Chiquita in Argentina, seashore saltgrass (*Distichlis spicata*), a clonally spreading grass, and the herb alkali weed (*Cressa truxillensis*) are characteristic of disturbed ground. Pickleweed (*Salicornia perennis*; formerly *Sarcocornia perennis*) establishes itself later and, lastly, denseflower cordgrass (*Sporobolus montevidensis*; formerly known as *Spartina densiflora*) arrives. Plant establishment is affected by grazing of young shoots by crabs and by wild guinea pigs. Researchers created bare patches in the salt marsh and used cages to exclude herbivores and prevent colonisation by clonal plants to investigate the separate and combined effects of grazing and vegetative spread of nearby plants. After 6 years, they harvested the plants on their experimental plots and recorded their biomass (Figure 2.6).

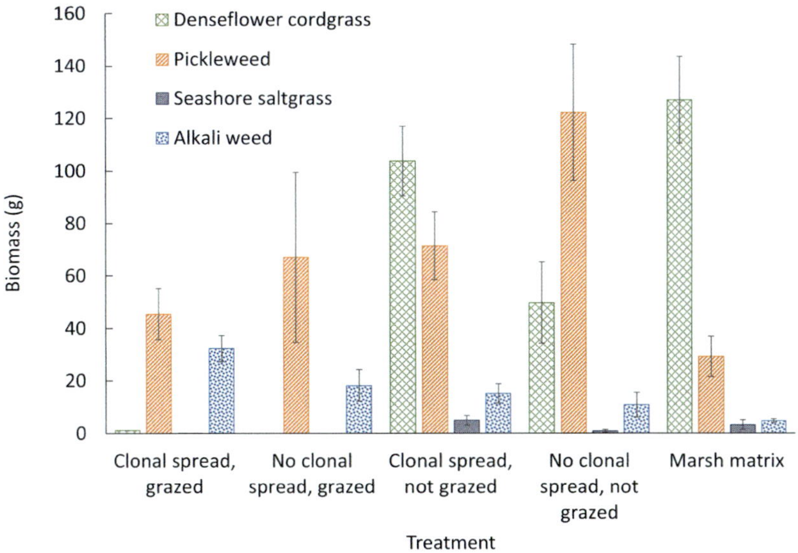

Figure 2.6 Biomass of salt marsh plants after 6 years of growth showing effects of grazing and clonal spread. Bars show means with ±1 SE. The biomass of each species on adjacent undisturbed salt marsh is shown as 'marsh matrix.' (Adapted from Daleo et al., 2014. Reproduced with permission, copyright 2014 Springer-Verlag Berlin Heidelberg.)

Question 2.6.1: The researchers included measurements of the undisturbed marsh matrix in addition to their treatments to check for experimental effects on the controls ('caging' effects). Which experimental treatment is the equivalent of the marsh matrix results?

A. Clonal spread, grazed
B. No clonal spread, grazed
C. Clonal spread, not grazed
D. No clonal spread, not grazed

Question 2.6.2: What is the most likely reason that the marsh matrix treatment is different from its equivalent control? (No more than 40 words are needed.)

Question 2.6.3: Which activity had the biggest impact on community composition after 6 years of salt marsh gap regeneration?

A. Clonal spread.
B. Grazing.
C. Both are equally important.

Queston 2.6.4: Which species do the herbivores prefer to eat?

A. Seashore saltgrass
B. Alkali weed
C. Pickleweed
D. Denseflower cordgrass

Question 2.6.5: Using the Connell and Slatyer (1977) model of succession, what impact does grazing have on this salt marsh system?

A. Facilitation
B. Inhibition
C. Tolerance

Exercise 2.7: Rocky shore succession

In 1974, as part of an undergraduate field course, Menge and Lubchenko set up an experiment in a sheltered cove in Nahant, Massachusetts. The dominant seaweed knotted wrack (*Ascophyllum nodosum*) and common barnacles (*Semibalanus balanoides*), which can provide important refuges for colonising species, were stripped from a series of plots. The recolonisation of these areas was documented until 1979. Anecdotal observations in 2009 showed that the plots had not been recolonised by knotted wrack even 30 years later and resulted in detailed resurveys in 2011 and 2013.

The original experiment had three treatments: knotted wrack retained (+), barnacles retained (+); knotted wrack removed (–), barnacles retained (+); knotted wrack removed (–), barnacles removed (–). Figure 2.7a shows some of the experimental results.

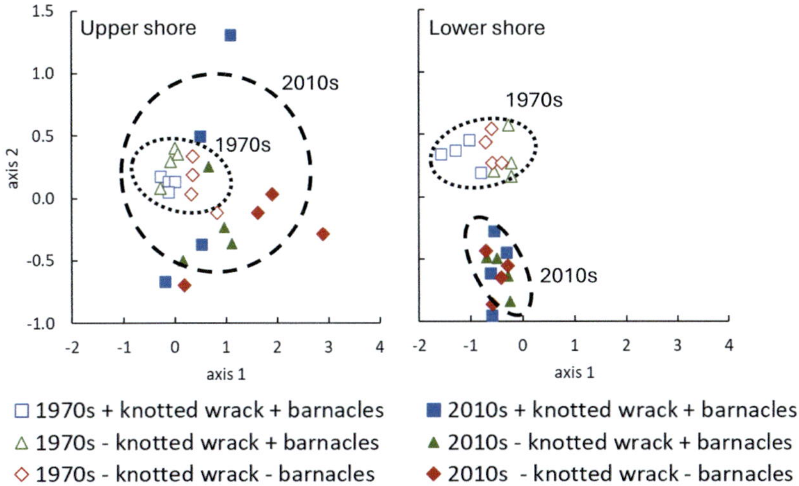

Figure 2.7a Non-metric multidimensional scaling analysis (see Preface for more details on this type of analysis) shows differences in communities on the upper shore (left) and lower shore (right) in the 1970s and 2010s. Ellipses show 95% CIs surrounding identified groupings; k = 3; stress = 0.09. (Adapted from Menge et al., 2017. Reproduced under the Creative Commons CC BY License, copyright 2017 Menge et al.)

Question 2.7.1: Which of the following statements summarises the key finding of the NMDS analysis?

A. In the 2010s, the *upper* shore was more *homogeneous* compared to the lower shore.
B. In the 2010s, the *upper* shore was more *heterogeneous* compared to the lower shore.
C. On the *lower* shore there was a greater difference between the communities in the 1970s and the 2010s than was seen on the *upper* shore between the two sampling periods.

Figure 2.7b shows percent cover of brown algae and knotted wrack before the experimental manipulation and 30 years later.

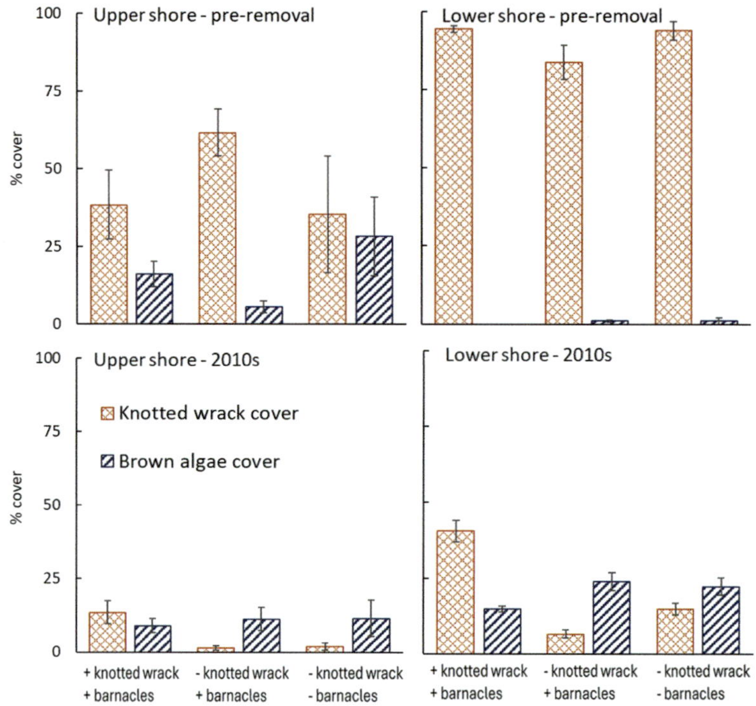

Figure 2.7b Percent cover of knotted wrack and brown algae (*Fucus*; another genus of seaweed) on the upper and lower shores before the experiment began and in the 2010s. (Adapted from Menge et al., 2017. Reproduced under the Creative Commons CC BY License, copyright 2017 Menge et al.)

Question 2.7.2: Which of the treatments was the control in this experiment?

A. + knotted wrack, + barnacles
B. – knotted wrack, + barnacles
C. – knotted wrack, – barnacles

Question 2.7.3: The large reduction in knotted wrack recorded on the lower shore before the start of the experiment and the 2010s was due to the removal of this species and its very slow recolonisation of the plots. True or false?

Question 2.7.4: On the upper shore, is there any evidence that barnacles facilitated the reestablishment of knotted wrack and brown algae?

Question 2.7.5: Does knotted wrack out-compete brown algae? Explain your reasoning. (No more than 30 words are needed.)

Question 2.7.6: What description could be used for this long-term change in a community? (Only three words are needed.)

Exercise 2.8: Invasive species and grassland succession

We hear many reports of invasive species and the issues they can cause, but most non-native species arriving in a new setting, even if they do establish, do not become problematic. Understanding which species will invade communities and how they are able to displace native species is an important area of research so that we can focus our action on the most relevant species.

Pearson and co-workers examined this issue in grasslands in western North America where cheatgrass *Bromus tectorum* has spread across large areas. The usual successional pattern following disturbance is the establishment of annual species which persist for a few years before being out-competed by perennial species which form the climax community. To test their ideas, the researchers removed all vegetation from plots and created 'starting communities' by seeding with *annuals*: either native, naturalised or invasive. Control plots were left unseeded. After one growing season, native *perennial* species were seeded into the plots. The following growing season, the numbers of individual perennial plants which had established in the annual starting communities were recorded.

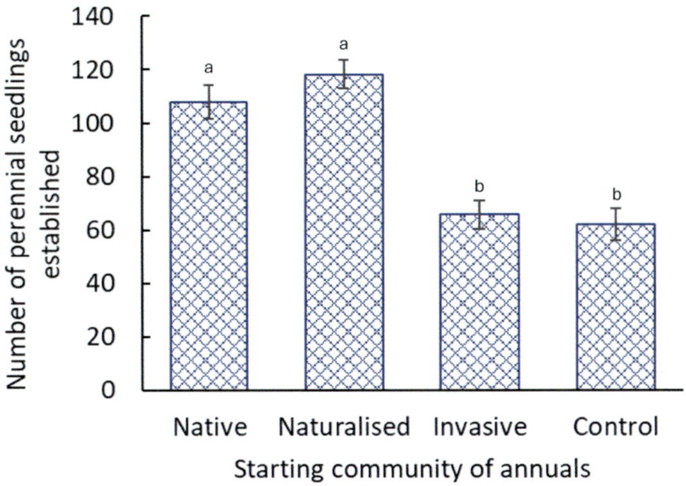

Figure 2.8a Number of perennial seedlings successfully establishing in starting communities of different types of annuals. Bars show mean ± 1 SE; bars sharing a letter are not significantly different from each other. (Adapted from Pearson et al., 2024. Reproduced under the CCO Public Domain Dedication, copyright Pearson et al., 2024.)

Question 2.8.1: Select the two correct statements about Figure 2.8a from the following list:

A. The highest establishment of perennial species was in the naturalised species starting community.
B. Perennial seedlings established equally well in the native and naturalised species starting communities.

C. Fewer individuals of the annual invasive species were able to establish in the starting community, so there was more space available for the perennial species to establish.
D. Significantly fewer perennial plants established in the invasive annual starting community than in the native starting community.
E. There were fewer invasive species in the starting community than in native or naturalised species communities.

Question 2.8.2: Which model of succession is suggested by the *native* annuals and perennials in these grasslands? (One word only!)

Question 2.8.3: Which model of succession is suggested by the *invasive* annuals and perennials in these grasslands? (One word only!)

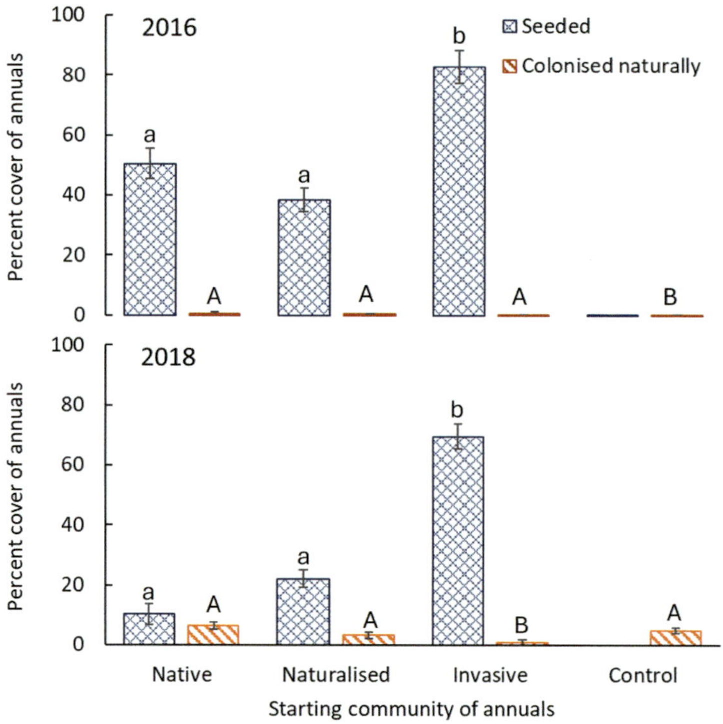

Figure 2.8b Percent cover of starting communities after 1 year (2016) and 3 years (2018) establishment. Bars show mean ± 1 SE. Bars sharing the same letter are not significantly different from each other. Bars for 'seeded' percent covers (shown with lower case) were not compared with 'colonised naturally' percent covers (shown with upper case). (Adapted from Pearson et al., 2024. Reproduced under the CCO Public Domain Dedication, copyright Pearson et al., 2024.)

Question 2.8.4: Use Figures 2.8a and 2.8b to explain how invasive annuals are able to become the dominant species in these grasslands. Aim for 100 words to give a complete explanation.

3

Energy, matter and decomposition

Exercise 3.1: Salamanders in streams

An experiment manipulated the headwaters of a stream dominated by about 20 taxa of macroinvertebrates and predatory salamanders. Light is limiting in this forest stream, and most energy enters the system as terrestrial leaf detritus. Macroinvertebrates and salamanders prey on small and large invertebrates (described as 'primary consumers'). After 2 years of measurements of the unmanipulated system, researchers added N and P to the stream for 4 years to stimulate the producer community and measured the effects on the rest of the food web.

 DOI: 10.1201/9781003644095-3

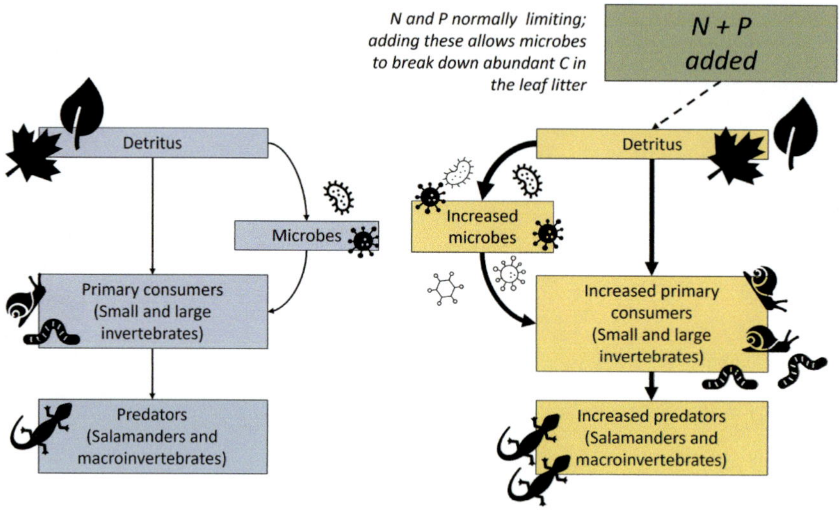

Figure 3.1a Schematic to show expected outcome. The left-hand side (blue boxes) shows the functioning of the system prior to manipulation. The right-hand side (yellow boxes) shows the anticipated results following the addition of N and P. The direction of arrows shows the movement of energy through the food web; thicker arrows indicate increased production. (Adapted from Davis et al., 2010. Reproduced with permission.)

Question 3.1.1: Using the details in Figure 3.1a, which pair of graphs (A, B, C or D) in Figure 3.1b represents the researchers' *expected* results? Y-axes for primary consumers and predators show production, both are measured in mg AFDM.m^{-2}.yr^{-1}; AFDM is ash-free dry mass, a measurement of productivity. Bars show means.

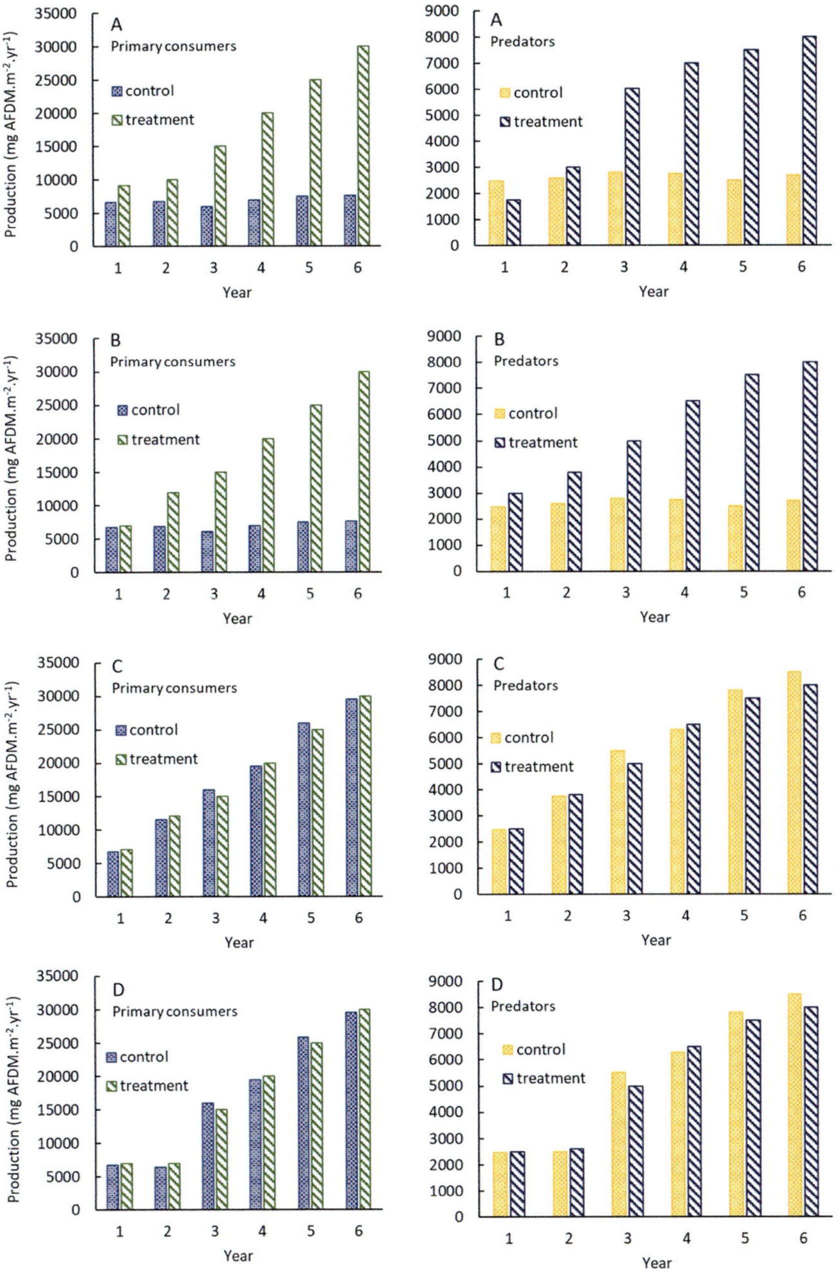
A
Primary consumers
control
treatment
Production (mg AFDM.m-2.yr-1)
Year
A
Predators
control
treatment
B
Primary consumers
control
treatment
B
Predators
control
treatment
C
Primary consumers
control
treatment
C
Predators
control
treatment
D
Primary consumers
control
treatment
D
Predators
control
treatment

Figure 3.1b

Figure 3.1c shows the actual results of the experiment to investigate productivity of primary consumers and predators following addition of N and P.

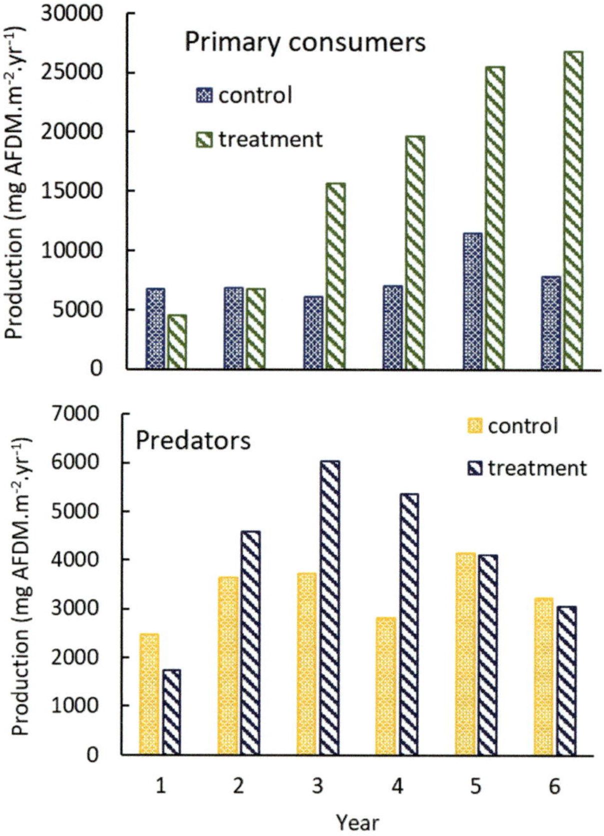

Figure 3.1c Production of primary consumers (top) and predators (bottom) after 4 years of additions of N and P. Bars show means. (Adapted from Davis et al., 2010. Reproduced with permission.)

Question 3.1.2: After the addition of N and P, production in primary consumers _____.

A. Decreased
B. Increased
C. Stayed the same
D. Initially increased then decreased
E. Initially decreased then increased
F. Initially increased then returned to the same levels as the controls
G. Initially decreased then returned to the same levels as the controls

Question 3.1.3: After the addition of N and P, production in predators _____.

A. Decreased
B. Increased
C. Stayed the same
D. Initially increased then decreased
E. Initially decreased then increased
F. Initially increased then returned to the same levels as the controls
G. Initially decreased then returned to the same levels as the controls

Question 3.1.4: The researchers reported that caddis flies in the genus *Pycnopsyche* (large-bodied caddis flies which build themselves cases out of small stones) increasingly dominated the primary consumer community over the last 4 years of the experiment. How might this help to explain the results?

A. The weight of the stone case increased the measured biomass.
B. *Pycnopsyche* spp. competed with predatory species.
C. Enrichment of this system allowed fish to spend more time in the formerly low productivity waters and they assumed the role of apex predators.
D. They are less vulnerable to predation.

Question 3.1.5: Salamanders are a key predator in this community; they are described as 'gape-limited' predators. How does this help to explain the results of this experiment? (No more than 50 words are needed.)

Question 3.1.6: The process where increased productivity at the base of a food web doesn't transmit to higher levels of the food web is known as 'trophic decoupling.' What are the two most likely destinations for the additional energy in this system that is not being consumed by predators? How does the energy move there? (Only 30 words are needed.)

Exercise 3.2: Soils over time in Hawaii

The island chain of Hawaii formed as the Pacific tectonic plate drifted over a stationary volcanic hotspot. This resulted in islands of very different ages and thus soils of different ages. Vitousek and Farrington (1997) published data from a 2-year study conducted in forests in Hawaii. They added N (100 $kg.ha^{-1}.yr^{-1}$) and P (100 $kg.ha^{-1}.yr^{-1}$) in factorial combinations to forest plots and measured above-ground tree growth (diameter stem increment, $mm.year^{-1}$). The study took place on three islands that were 300, 20,000 and 4.1 million years old. Figure 3.2 shows some of the findings.

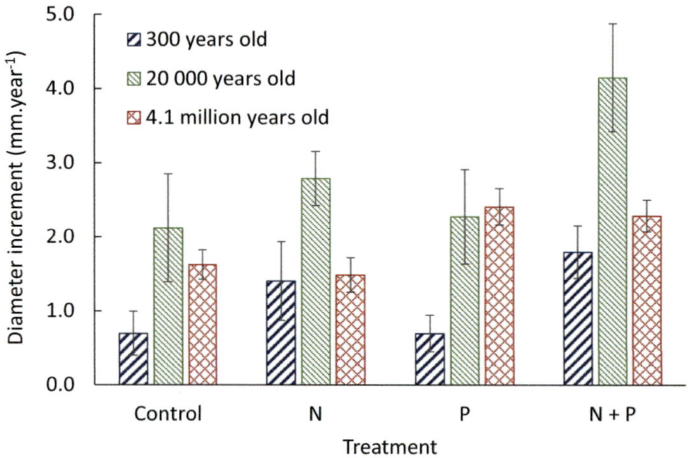

Figure 3.2 Effect of adding N, P and N+P on the growth increment of trees. Bars show mean ± 1 SE. (Adapted from Vitousek and Farrington, 1997. Reproduced with permission, copyright 1997 Kluwer Academic Publishers.)

Question 3.2.1: Why did Vitousek and Farrington add N and P to their study plots? Why did they add the nutrients separately and together? (No more than 25 words are needed.)

Question 3.2.2: What is the main route for P to enter soils?

A. Fixing from the atmosphere by legumes
B. Weathering of rocks
C. Atmospheric deposition
D. Rainfall
E. Lightning

Question 3.2.3: What is the main route for N to enter soils?

A. Fixing from the atmosphere by legumes
B. Weathering of rocks
C. Atmospheric deposition
D. Rainfall
E. Lightning

Question 3.2.4: Select the following statements which best describe Figure 3.2. (There are four correct answers.)

A. On the 300-year-old soils, adding N+P resulted in the greatest increase in growth, followed by adding just N.
B. On the 300-year-old soils, adding N and adding N+P gave the same result.

C. On the 300-year-old soils, there was no significant difference between adding P and the controls with no P or N added.
D. On the 20,000-year-old soils, adding N+P was the only treatment to increase growth of the trees.
E. On the 20,000-year-old soils, adding N+P and adding just N increased growth, but there was no difference between the P treatment and control.
F. On the 4-million-year-old soils, adding P and N+P increased growth of trees, but adding N alone had no effect.
G. On the 4-million-year-old soils, adding P had the greatest effect on tree growth, followed by adding N+P.

Question 3.2.5: Which nutrients are limiting on each island and why is this likely to have occurred? (For a complete answer, aim for around 100 words.)

Exercise 3.3: Coral spawning and nutrients

Many species of coral spawn synchronously, releasing vast numbers of eggs and sperm which float to the sea surface. Some eggs are fertilised and the larvae sink to the seafloor a few days later. However, the vast majority of eggs do not get fertilised and die after a few days; this creates a short-lived but large input of accessible, nutrient-rich material into coral reef systems. Wild et al. (2004) investigated a minor coral spawning event in a lagoon on Heron Island, part of the Great Barrier Reef, in 2001. Figure 3.3 shows some of their results.

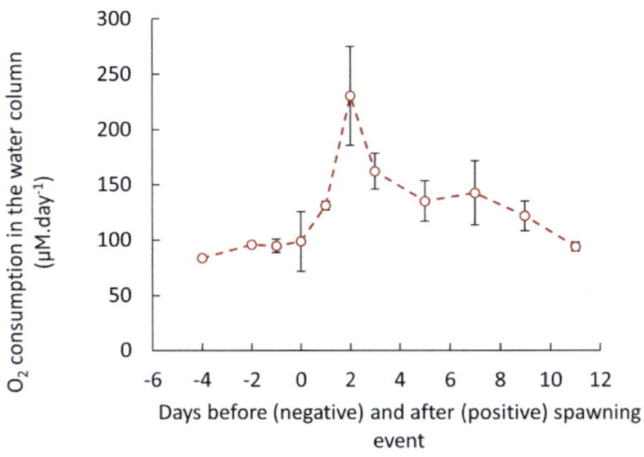

Figure 3.3 Sedimentary O_2 consumption before, during and after a coral spawning event. Points represent mean ± 1 SD. (Adapted from Wild et al., 2007. With permission, copyright Inter Research.)

Question 3.3.1: Wild et al. wanted to quantify the impact of the addition of nutrients to the coral reef sediment. Why did they measure O_2 consumption?

A. O_2 is found in high quantities in the coral eggs and sperm, so a peak in O_2 just after coral spawning shows that these have arrived in the benthic sediment.
B. This told them about the respiration of organisms in the benthic sediment community.
C. Phytoplankton on the surface of the sediment are nutrient limited and respond to the flush of nutrients with a large increase in photosynthesis.
D. A peak in O_2 production would tell them about the arrival of other organisms, particularly filter-feeding fish, which arrive to take advantage of the input of coral eggs into the system.

Question 3.3.2: What does the peak in O_2 consumption directly after the coral spawning event tell us about the nutrient status of the sediment and the organisms that live there? (Only 20 words are needed.)

Question 3.3.3: The graph shows a sharp peak in O_2 consumption straight after the coral spawning event, followed by a gradual decline to day 11. What might explain this gradual decrease rather than O_2 consumption dropping back to its original levels within 1–2 days? (No more than 60 words are needed.)

Exercise 3.4: Decomposition by woodlice

Two or six woodlice *Oniscus asellus*, a small, common arthropod particularly found in rotting wood, were added to microcosms containing 1 g of leaf litter; a control microcosm contained no woodlice. Figure 3.4 shows results of various measurements taken by the researchers.

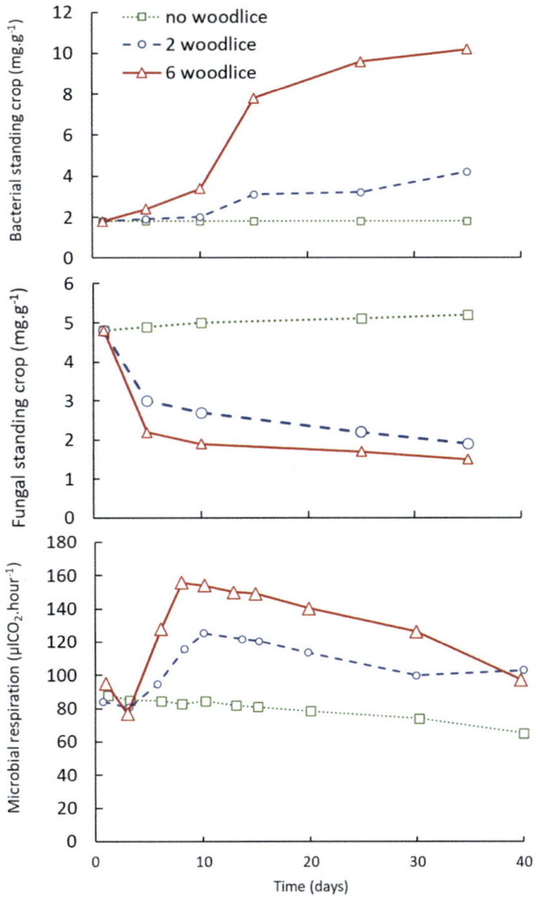

Figure 3.4 Bacterial standing crop over 35 days (top), fungal standing crop over 35 days (middle) and microbial respiration over 40 days (bottom). Points represent means. (Adapted from Hanlon and Anderson, 1980. Reproduced with permission, copyright 1980 Elsevier Ltd.)

Question 3.4.1: Which of the following statements is the most likely reason for the decrease in the fungal standing crop when woodlice were present?

A. Bacteria secrete compounds that inhibit fungal growth.
B. Fungi are less good at competing against woodlice and ran out of food.
C. Fungi are also eaten by woodlice.

Question 3.4.2: Why did bacterial standing crop increase when there were six woodlice present compared with only two or no woodlice? (No more than 60 words are needed.)

Exercise 3.5: Home-field advantage for decomposing litter?

There has been considerable debate over the key factors affecting decomposition rates of plant litter. It has been observed that litter lying beneath its own tree decomposes faster than that under other trees, suggesting that local microbial communities are specialised in the litter they most commonly encounter. This is known as 'home-field advantage.' Alternatively, local environmental or climatic factors acting on decomposer communities may be an important influence.

Shigyo et al. (2024) investigated the relative importance of site and soil in an experiment in temperate forests in central Japan. They used plant litter from northern Japanese hemlock (*Tsuga diversifolia*), an upland species, and Japanese beech (*Fagus japonica*), found in lowland areas. Litter from both species was buried at upland and lowland sites. To untangle the role of environment and soil, they also translocated soil between sites. Thus, litter from both trees was buried in soil from upland and lowland sites at both the upland and lowland sites. Litter mass loss was determined after 527 days.

Question 3.5.1: If the home-field advantage hypothesis is correct, which of the following options would you expect to see? (More than one answer is correct.) Greater decomposition for:

A. Hemlock litter in upland soil on upland sites.
B. Hemlock litter in upland soil on lowland sites.
C. Hemlock litter in lowland soil on upland sites.
D. Hemlock litter in lowland soil on lowland sites.
E. Beech litter in upland soil on upland sites.
F. Beech litter in upland soil on lowland sites.
G. Beech litter in lowland soil on upland sites.
H. Beech litter in lowland soil on lowland sites.

Question 3.5.2: If the environment is the key driver of decomposition, which of the following options would you expect to see? (More than one answer is correct.)

Greater decomposition for:

A. Hemlock litter in upland soil on upland sites.
B. Hemlock litter in upland soil on lowland sites.
C. Hemlock litter in lowland soil on upland sites.
D. Hemlock litter in lowland soil on lowland sites.
E. Beech litter in upland soil on upland sites.
F. Beech litter in upland soil on lowland sites.
G. Beech litter in lowland soil on upland sites.
H. Beech litter in lowland soil on lowland sites.

Figure 3.5 shows some of the results from the Shigyo et al. (2024) study.

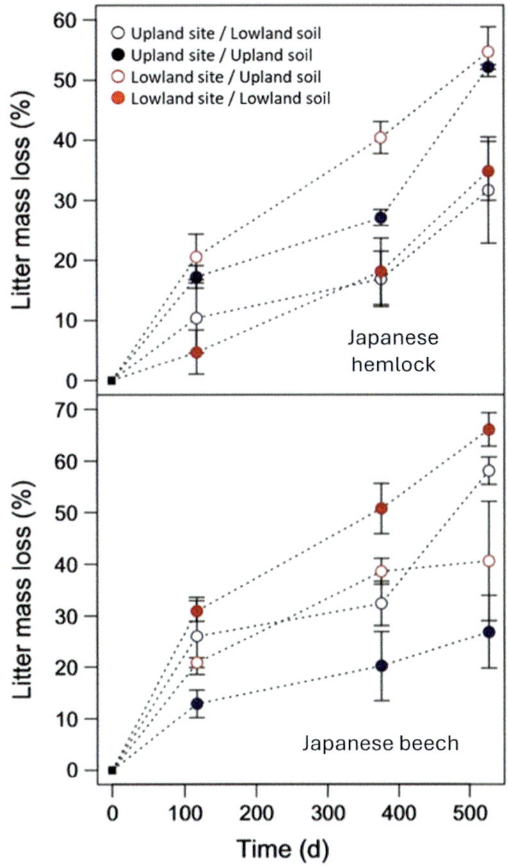

Figure 3.5 Percentage litter mass loss of hemlock and beech over 527 days at upland and lowland sites and in upland and lowland soil. Points show mean percentage mass loss with ±1 SD. (From Shigyo et al., 2024. Reproduced under the Creative Commons CC BY licence, copyright 2024 Shigyo et al., 2024; New Phytologist, copyright 2024 New Phytologist Foundation.)

Question 3.5.3: Based on these findings, which factor is more important in determining rates of decomposition?

A. Home-field advantage
B. Environment

Exercise 3.6: Plant litter and termites

In Hluhluwe-iMfolozi Park in South Africa, termites play a large role in moving and processing plant litter. Despite comprising a biomass that can exceed that of large herbivores, their role is often overlooked. Researchers used long-term exclosures to prevent access by large herbivores (white rhino, buffalo, zebra, impala and warthog; 'ungrazed') and compared these with similar-sized areas which large grazers could access. On the ground in each exclosure, they placed litter bags with a mesh size that only allowed organisms the size of termites (or smaller) to enter and remove litter; they used the same, high-quality litter in all litter bags. They also noted that at the wet sites the vegetation contained high concentrations of phenolics and lignin (Figure 3.6).

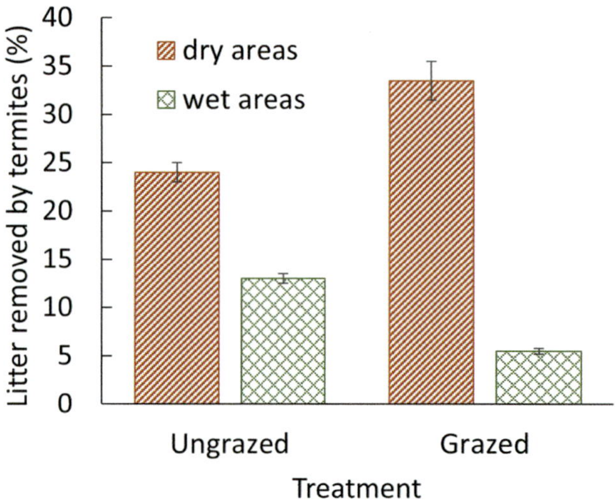

Figure 3.6 Percentage litter in litter bags removed by termites in wet and dry areas which were grazed or ungrazed. Bars show mean ± 1 SE. (Adapted from Buitenwerf et al., 2011. Reproduced with permission, copyright Cambridge University Press 2011.)

Question 3.6.1: Why is it important to consider the concentrations of phenolics and lignin in vegetation surrounding the litter bags? (No more than 50 words are needed.)

Question 3.6.2: Which of the following two options are the most likely reasons that more litter was removed by termites in the dry, grazed exclosures, compared with the dry, ungrazed ones?

A. Termites prefer to forage where the vegetation is short.
B. Different vegetation was growing in the grazed and ungrazed plots.
C. Termites don't forage far from their nests; by chance there were more termite nests inside and close to the grazed plots.
D. Large herbivores drop bitten-off pieces of vegetation as they chew.
E. There is undigested grass in the dung of large herbivores.

Question 3.6.3: How would you describe the interaction between large herbivores and termites on sites with wet soil? How does this change on sites with dry soil? (No more than 40 words are needed.)

Exercise 3.7: Whales and nutrient cycling

Despite differences in feeding style, all cetaceans tend to urinate and defecate near the sea surface; this has the potential to move nutrients considerable distances and create nutrient 'hotspots.' A team of researchers from La Rochelle and Nantes in France modelled data on the contribution made by whale faeces to nutrient cycling in various parts of the oceans, focusing on two major nutrients (N and P) and six trace elements (Fe, Cu, Mn, Se, Zn and Co). They analysed their data using principal component analysis (PCA), another type of multivariate analysis (see the Preface for more details on how to interpret this type of output).

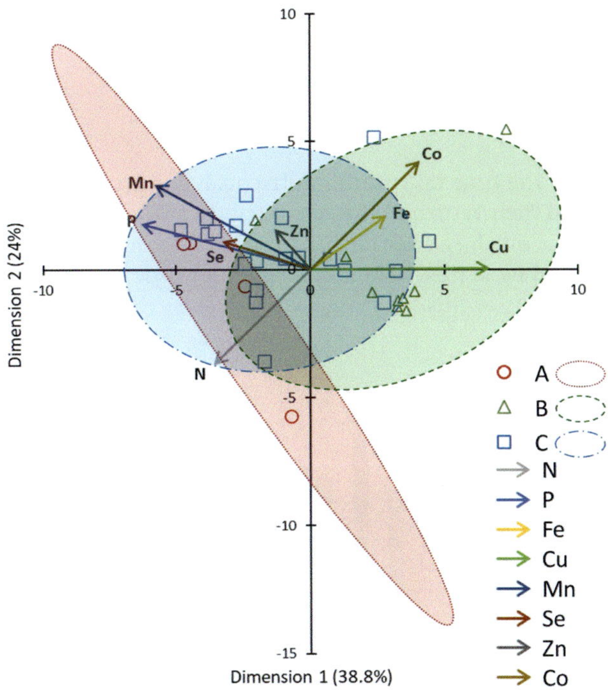

Figure 3.7a Results of a PCA for nutrient composition of faeces released by different groups of cetaceans. Each point represents one species. Species are grouped by type: baleen whales, small cetaceans (dolphins and porpoises smaller than 4 m long) and deep divers (sperm, beaked and pilot whales; and Risso's dolphin). (From Gilbert et al., 2023. Reproduced under the Creative Commons CC BY licence, copyright 2023 Gilbert et al.)

Question 3.7.1: The coloured ellipses (labelled A, B, C) in Figure 3.7a represent different groups of cetaceans. Use the relationship between the ellipses and the eigenvalues to help work out which group of cetaceans (baleen whales, small cetaceans, deep divers) is represented by each letter. The following bullet points give some extra information, but you need to have a basic knowledge of the types of food eaten by each group.

- Cephalopods and crustaceans use Cu-containing haemocyanin as a respiratory pigment.
- Bony fish contain high levels of P.
- Krill and crustaceans contain high levels of N.

Question 3.7.2: How much of the variation in the dataset is explained by the two-dimensional analysis?

A. 14.4%
B. 24.0%
C. 38.8%
D. 62.8%
E. 95.0%
F. 100.0%

Question 3.7.3: The time taken from nutrients being assimilated into organic compounds and their return to the ecosystem as available nutrients (i.e. their turnover time) is much shorter for baleen whales than for small cetaceans or deep divers. (i) Why is this? (No more than 30 words are needed.) (ii) What are the implications of this for primary producers? (No more than 20 words are needed.)

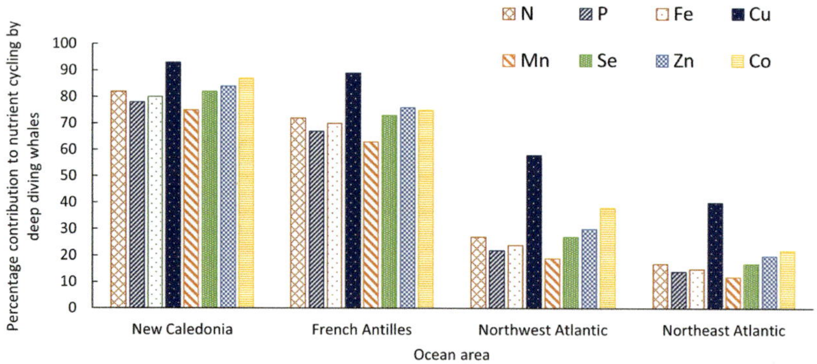

Figure 3.7b The proportion of nutrients cycled by deep-diving cetaceans in tropical areas (New Caledonia and French Antilles) and the North Atlantic. For the tropical areas, the remaining proportion of nutrients is cycled by small cetaceans; for the North Atlantic, the remaining proportions are split between baleen whales and small cetaceans. (Adapted from Gilbert et al., 2023. Reproduced under the Creative Commons CC BY licence, copyright 2023 Gilbert et al.)

Question 3.7.4: Figure 3.7b shows that the proportion of nutrients not cycled by deep-diving whales differs between the North Atlantic and tropical areas. Why is this important in relation to nutrient cycling?

A. Baleen whales cannot feed in nutrient-poor tropical waters.
B. Small cetaceans tend to be out-competed by deep-diving cetaceans in tropical areas; the reverse is true in temperate regions.
C. In nutrient-poor tropical waters, deep-diving cetaceans recycle nutrients already present (autochthony).
D. In nutrient-poor tropical waters, deep-diving cetaceans provide a source of allochthonous (external) nutrients.

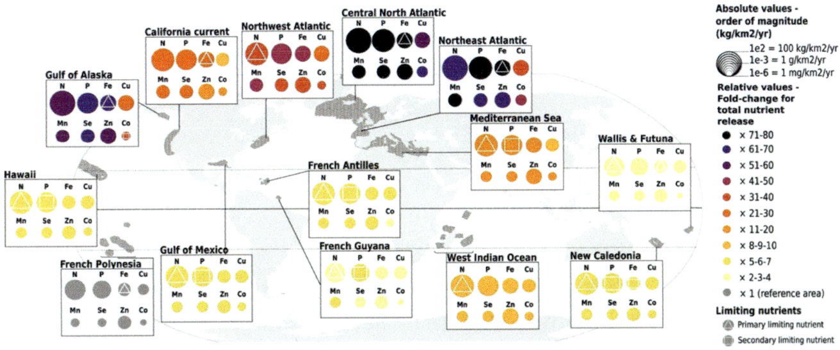

Figure 3.7c For each marine area, the nutrients released by cetaceans in faeces are shown. The size of each circle shows the absolute amounts of nutrients (the smallest circle shows 1 $mg.km^{-2}.yr^{-1}$; largest circle shows 100 $kg.km^{-2}.yr^{-1}$). Colour-coding shows the release of nutrients relative to the reference area, French Polynesia (the area with the lowest absolute values). Triangles within circles indicate primary limiting nutrients, and squares show secondary limiting nutrients. (From Gilbert et al., 2023. Reproduced under the Creative Commons CC BY licence, copyright 2023 Gilbert et al.)

Question 3.7.5: There is a lot going on in Figure 3.7c, but we will focus on some key points. In the tropics, which two nutrients are usually limiting?

A. N
B. P
C. Fe
D. Cu
E. Mn
F. Se
G. Zn
H. Co

Question 3.7.6: In temperate regions, which nutrient is usually limiting?

A. N
B. P
C. Fe
D. Cu
E. Mn
F. Se
G. Zn
H. Co

Question 3.7.7: In which geographic area do cetaceans release most nutrients?

A. Tropics
B. Sub-tropics
C. Northern temperate regions
D. Southern temperate regions

Exercise 3.8: Nutrients, mangroves and seabirds

Stable isotope analysis allows us to infer the food sources of organisms. Most N is ^{14}N (it has 7 protons and 7 neutrons); a small proportion has 7 protons and 8 neutrons (^{15}N). When organisms synthesise protein, they preferentially take up ^{15}N. This means that a consumer is enriched in $\delta^{15}N$ by approximately 3.4‰ (parts per thousand) compared with the food it has eaten. Their faeces will also have higher ^{15}N than the background level. ^{15}N analysis can reveal the horizontal transfer of nutrients between organisms within an ecosystem.

Mangroves are found in tropical and subtropical areas and provide important ecosystem services such as protecting coastal communities from storms. They are also important for biodiversity, local livelihoods and sequester large amounts of C. Researchers investigated the nutrient dynamics of mangrove forests on Aldabra Atoll in the Seychelles. They used ^{15}N analysis to investigate whether seabirds transported nutrients into mangrove forests, by comparing areas used as nest sites and those without birds.

Table 3.8a N and $\delta^{15}N$ content of guano from red-footed boobies and frigatebirds on Aldabra Atoll

	N content (%) of guano (Mean ± 1 SD)	$\delta^{15}N$ (‰) of guano (Mean ± 1 SD)
Red-footed booby (n = 4)	10.1 ± 2.92	10.8 ± 2.71
Frigatebird (n = 20)	10.9 ± 2.06	12.9 ± 3.63

Source: Appoo et al., 2024. Reproduced with permission, copyright 2024 Appoo et al.

Question 3.8.1: Choose the most likely explanation of the results in Table 3.8a.

A. The guano of the seabirds is depleted in ^{15}N relative to atmospheric N.
B. The guano of the seabirds is enriched in ^{15}N relative to atmospheric N.
C. There was a significant difference in concentration of N between red-footed boobies and frigatebirds.
D. There was a significant difference in concentration of $\delta^{15}N$ of guano of red-footed boobies and frigatebirds.

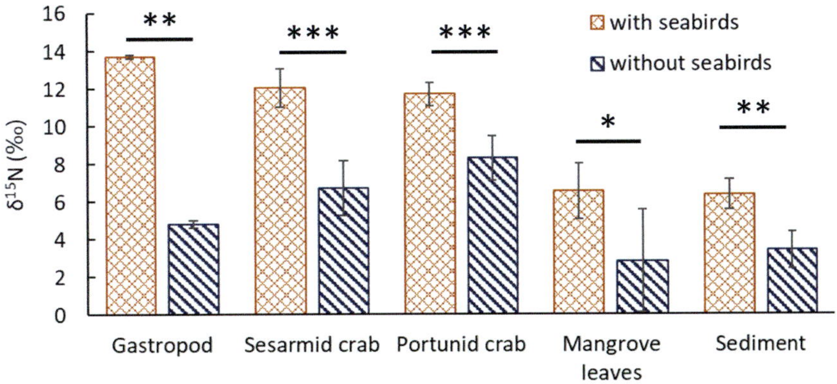

Figure 3.8 $\delta^{15}N$ content of species from environments with and without seabirds. Bars show mean ± 1 SD. Asterisks show the level of significant difference between samples taken from areas with and without seabirds: *$P < 0.05$; **$P < 0.01$; ***$P < 0.001$. (Adapted from Appoo et al., 2024. Reproduced with permission, copyright 2024 Appoo et al.)

Question 3.8.2: Using the values for areas without seabirds (see Figure 3.8), order the organisms by ascending trophic level, starting with the sediment.

Question 3.8.3: For each trophic group, the $\delta^{15}N$ in areas used by nesting seabirds is significantly higher than areas not used by seabirds. What is the ecological interpretation of this? (No more than 30 words are needed.)

Question 3.8.4: Additional work showed that C:N ratios were lower in seabird sites compared with non-seabird sites. Mangrove leaves had a mean N:P ratio of 11.7 ± 2.45 (SD). N:P ratios > 32 show that P is limiting. Based on this, which nutrient is likely to be limiting?

A. Carbon
B. Nitrogen
C. Phosphorus

Nutrient resorption in plants is a critical adaptation by plants growing in nutrient-deficient environments. Before leaves die and are shed, nutrients are resorbed thus conserving scarce nutrients. When nutrient limitation is reduced, nutrient resorption also decreases. However, the researchers found no difference in resorption efficiency by mangrove plants for N or P. Their results are shown in Table 3.8b.

Table 3.8b Resorption efficiency of mangrove plants in sites with and without seabirds

Resorption efficiency (Mean ± SD; %)	Seabird site	Non-seabird site
N	53.5 ± 13.1	59.5 ± 7.3
P	50.1 ± 25.2	59.2 ± 10.9

Question 3.8.5: Why might these mangrove plants still show strong nutrient resorption? (No more than 20 words are needed.)

4

Intraspecific competition and life histories

Exercise 4.1: Population cycles in Capitella capitata

Capitella capitata is a marine polychaete worm that lives in littoral or sublittoral mud. It occurs widely and is tolerant of polluted water. It is a generalist, suspension feeder, growing up to 1 cm long and living in burrows in the sediment (Keerthika et al., 2024). Chesney and Tenore (1985) investigated population cycles of this species under lab conditions, providing a constant amount of food each day throughout the experiment (Figure 4.1a).

 DOI: 10.1201/9781003644095-4

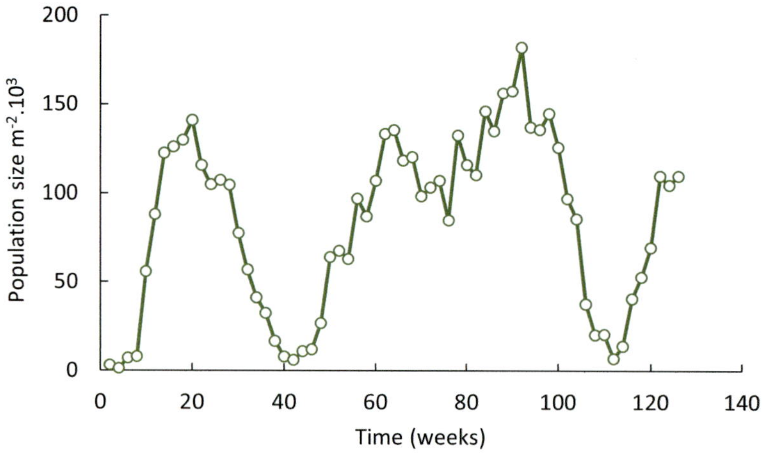

Figure 4.1a Population size of a single population of *Capitella capitata* recorded over 126 weeks. (Adapted from Chesney and Tenore, 1985. Reproduced with permission from Marine Ecology – Progress Series.)

Question 4.1.1: Between week 5 and week 20, what type of growth does this population show?

A. Logistic
B. Exponential

Question 4.1.2: The researchers calculated the carrying capacity (K) of this population to be around 80×10^3 m^{-2}. Why might the population repeatedly overshoot and then undershoot K? (No more than 50 words are needed.)

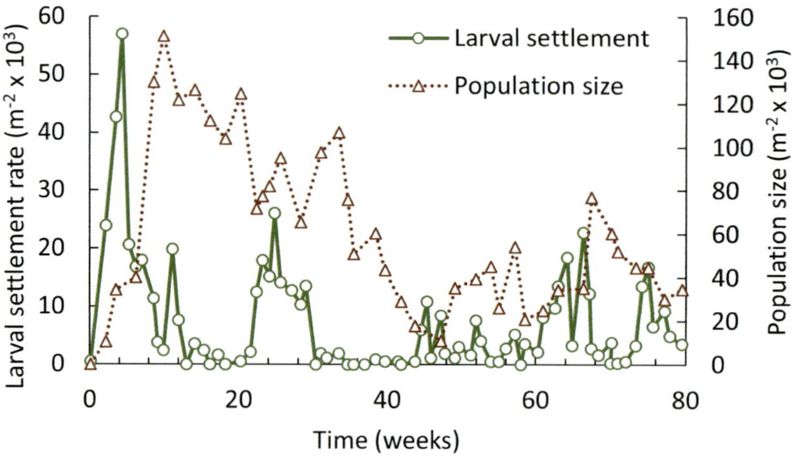

Figure 4.1b Larval settlement as a measure of reproductive rate and total population size over 80 weeks. (Adapted from Chesney and Tenore, 1985. Reproduced with permission from Marine Ecology – Progress Series.)

Question 4.1.3: Figure 4.1b shows a clear reduction in population size over 80 weeks. What is likely to be driving this decrease in population size? (No more than 20 words are needed.)

Exercise 4.2: Intraspecific competition over time in box tree moths

The box tree moth (*Cydalima perspectalis*) is native to Asia but was accidentally introduced in Europe where its larvae can cause extensive damage to box *Buxus* spp. trees and shrubs. It is a multivoltine species (i.e. there is more than one generation each year), so there is potential for intergenerational competition even though individuals from different generations may not be present simultaneously on the food plant.

Castagneyrol et al. (2021) investigated whether the presence of box tree moth larvae affected oviposition behaviour of subsequent generations and whether the presence of previous generations impacted the success of later generations (Figures 4.2a and 4.2b).

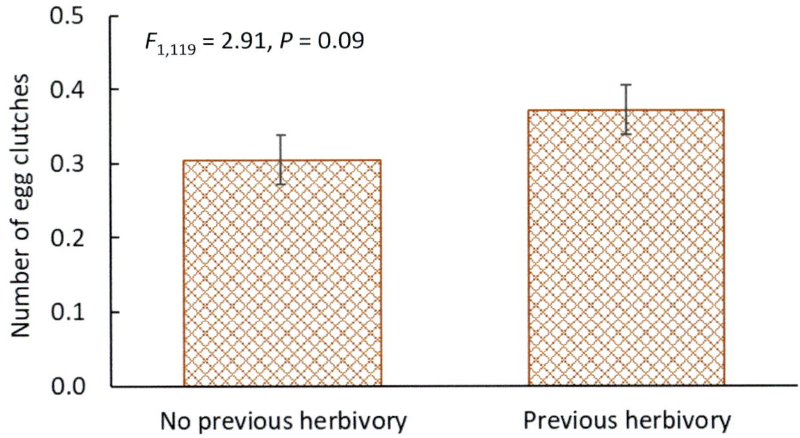

Figure 4.2a Bar chart showing the effect of earlier generations of moth larvae on egg laying. Bars show mean ± 1 SE. The graph also shows the results of a generalised least squares model to test for differences between previously grazed and ungrazed box trees. (Adapted from Castagneyrol et al., 2021. Reproduced under the Creative Commons BY CC 4.0 licence. Copyright 2021 Castagneyrol et al.)

Question 4.2.1: Did previous grazing make a difference to the number of egg clutches laid by later generations?

A. Yes, there were significantly fewer egg clutches on previously grazed box plants.

B. Yes, there were significantly more egg clutches on previously grazed box plants.
C. No, there was no significant difference between the number of egg clutches on previously grazed or ungrazed plants.

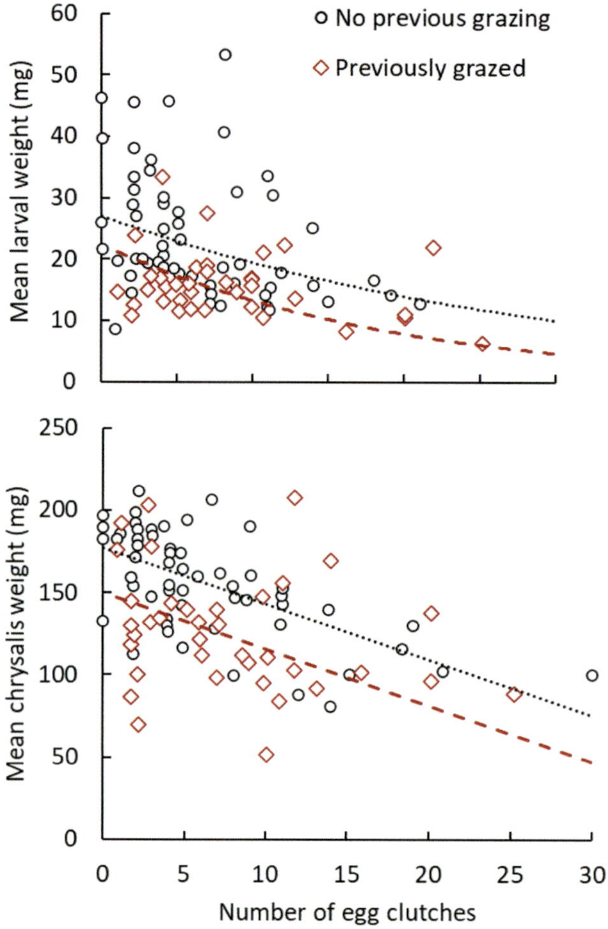

Figure 4.2b Effect of previous herbivory on larval weight and chrysalis weight of subsequent generations; points represent means. The lines are modelled fitted lines for each factor (grazing or no grazing). (Adapted from Castagneyrol et al., 2021. Reproduced under the Creative Commons BY CC 4.0 licence.)

Table 4.2 shows the corresponding results from the statistical tests.

Table 4.2 Results of analyses testing effect of prior herbivory on larval and chrysalis weight of subsequent generations

Response	Factor	df	F-value	P-value
Larvae	Number of egg clutches	1, 117	26.31	<0.001
	Previous herbivory	1, 117	20.30	<0.001
	Herbivory × egg clutches	1, 117	0.73	0.396
Chrysalis	Number of egg clutches	1, 100	33.74	<0.001
	Previous herbivory	1, 100	12.23	<0.001
	Herbivory × egg clutches	1, 100	3.14	0.079

Question 4.2.2: Work by Gripenberg et al. (2010) suggests that there is clear evidence for the 'preference–performance' hypothesis in herbivorous insects, i.e. that females will evolve to lay eggs on plants that will maximise the number of her offspring that survive. To do this, females must be able to recognise earlier grazing on plants. Can you suggest three signals that are likely to be used for this? Which of these are being used by box tree moths? (No more than 60 words are needed.)

Question 4.2.3: Caterpillars of box tree moths appear to be able to store or metabolise the highly toxic alkaloids present in box leaves. What effect might this have on female choice of individual host plants?

A. Larvae are less likely to be eaten by predators.
B. There is likely to be reduced competition with other species of grazing insect.
C. There is no requirement for strong camouflage in these moth larvae.
D. There isn't strong selection against females laying eggs on previously grazed leaves.

Question 4.2.4: Table 4.2 shows that the interaction for herbivory × eggs was not significant for larvae or chrysalis. But what might a significant interaction look like on the scatter plots? Choose the graph (A, B, C or D) in Figure 4.2c that represents how a significant interaction might look, had it occurred in this experiment.

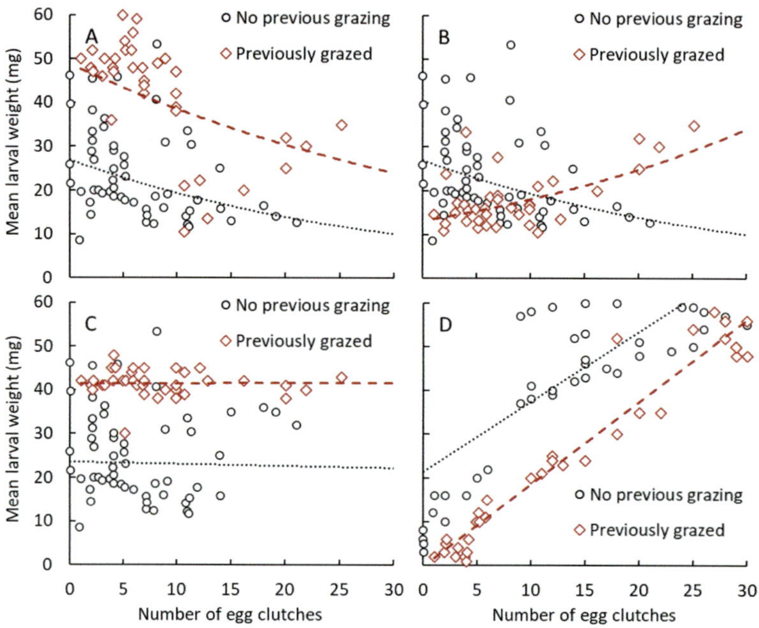

Figure 4.2c

Question 4.2.5: Having worked through the previous questions and information provided, is intergenerational, intraspecific competition occurring in box tree moths? Explain your reasoning. (No more than 50 words are needed.)

Exercise 4.3: Limpet density

The spoon limpet (*Scutellastra cochlear*; formerly known as *Patella cochlear*) lives on rocky shores. A population that was studied in South Africa showed densities of 125 (low), 355 (medium) and 1225 (high) individuals.m^{-2}. Limpets graze the algae that grow on the rocks and on the shells of other limpets, thus the area available for feeding remains ± constant regardless of the size and number of limpets.

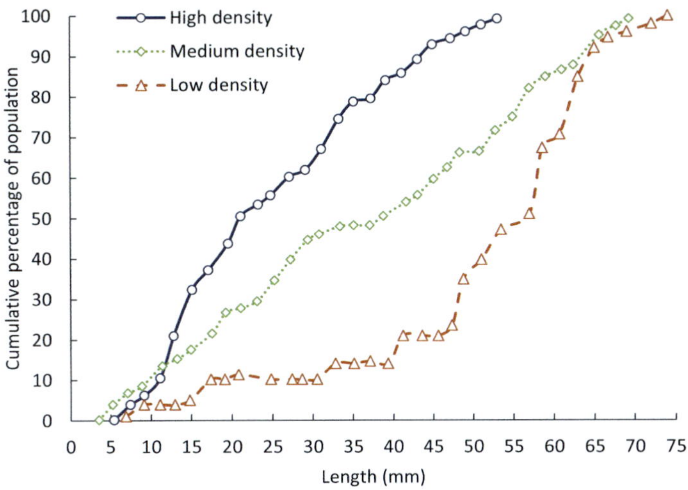

Figure 4.3a Cumulative proportion of limpet populations of different sizes in high-, medium- and low-density populations. (Adapted from Branch, 1975. Reproduced with permission of John Wiley & Sons – Journals.)

Question 4.3.1: Using Figure 4.3a, select the most appropriate statement to describe the findings.

A. Large limpets are more common in dense aggregations of limpets because they gain more protection against predatory birds; there is safety in numbers.
B. Growth of individual limpets is unaffected by population density.
C. As density increases, the maximum length of limpet shell decreases.
D. There is some evidence for the effects of interspecific competition and the role of predation in determining the size structure of limpet populations.

Figure 4.3b shows a subset of the data for all three population densities for individuals up to 20 mm long.

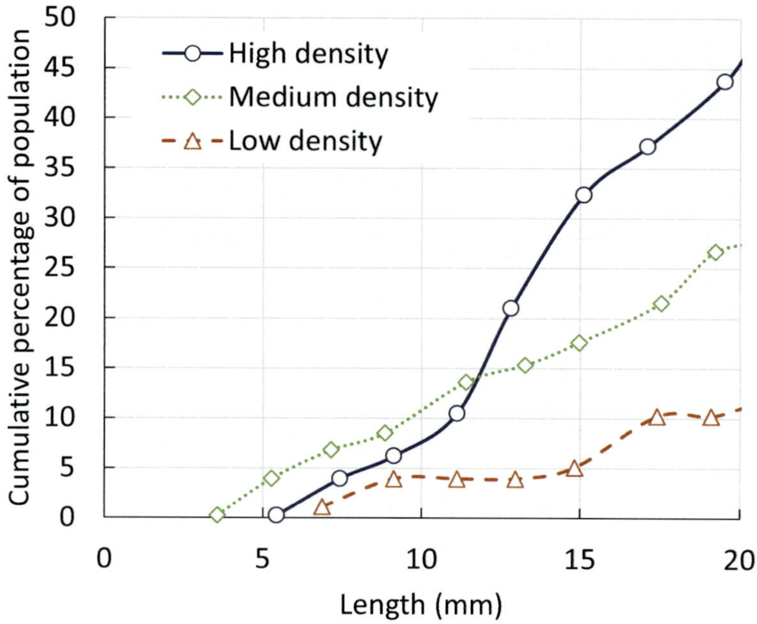

Figure 4.3b

Question 4.3.2: Using the more detailed graph in Figure 4.3b, what proportion of the low-density population is between 10 and 15 mm long? Choose the closest answer from the options given.

A. 1%
B. 4%
C. 5%
D. 17.5%
E. 22%
F. 33%

Figures reproduced with permission of John Wiley & Sons – Journals. Branch, G.M. 1975. Intraspecific competition in *Patella cochlear* Born. *Journal of Animal Ecology*, 44: 263-281; permission conveyed through Copyright Clearance Center, Inc.

Exercise 4.4: Trade-offs in life histories

New Mexico spadefoot toads (*Spea multiplicata*) spend most of their lives buried in sand in deserts, emerging only after rain to mate and lay eggs in pools. Pools generally don't persist for long in these environments. There are two morphs of tadpole in this species: one is carnivorous (feeding on fairy shrimps and other tadpoles), the other is omnivorous (feeding on algae and detritus). All tadpoles are initially the omnivorous morph. The trigger for development as a carnivore appears to be the abundance of shrimp in the pool—more shrimp result in more carnivores. Carnivorous tadpoles develop up to 50% faster than omnivorous morphs, but omnivorous tadpoles are able to build up greater fat reserves than carnivorous tadpoles (Figure 4.4a).

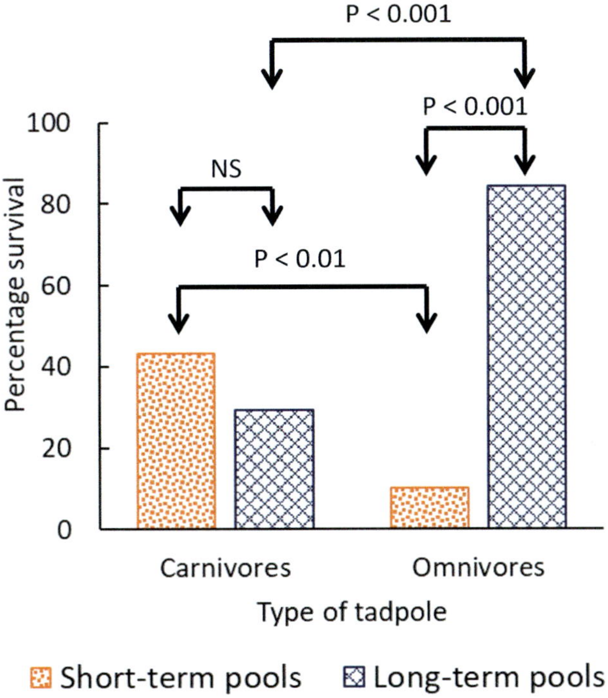

Figure 4.4a Survival of toads over 2 months after metamorphosis for both types of morphs in both types of pools. (Adapted from Pfennig, 1992. Reproduced with permission, copyright 1992 The Society for the Study of Evolution.)

Question 4.4.1: The bars with associated P-values are another way of showing the results of statistical tests, in this case a two-tailed Mann–Whitney *U*-test. If you had to display these results using a letter over each bar with different letters showing significantly different pairwise combinations, which of the graphs (A, B, C or D) in Figure 4.4b would be correct?

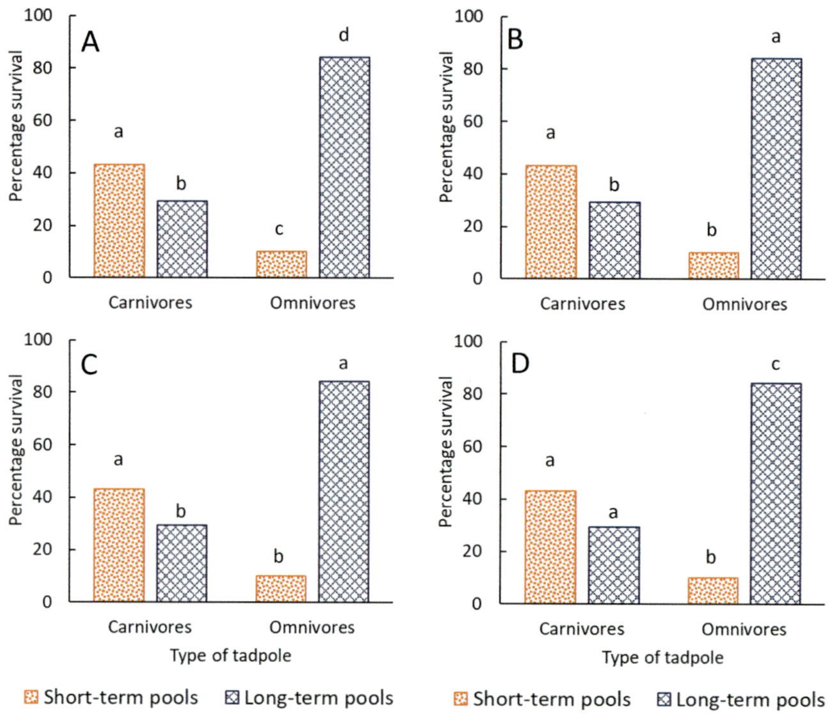

Figure 4.4b

Question 4.4.2: Select the true statement.

A. The most successful option is to be an omnivore.
B. The most successful option is to be a carnivore.
C. It is always better to be in a long-term pool.
D. It is always better to be in a short-term pool.
E. The worst option is to be omnivorous in short-term pools.

Question 4.4.3: In no more than 50 words, state why natural selection hasn't resulted in one type of life history for this species.

Pfennig (1992) manipulated the starting proportions of carnivores: omnivores. The results are shown in Figure 4.4c.

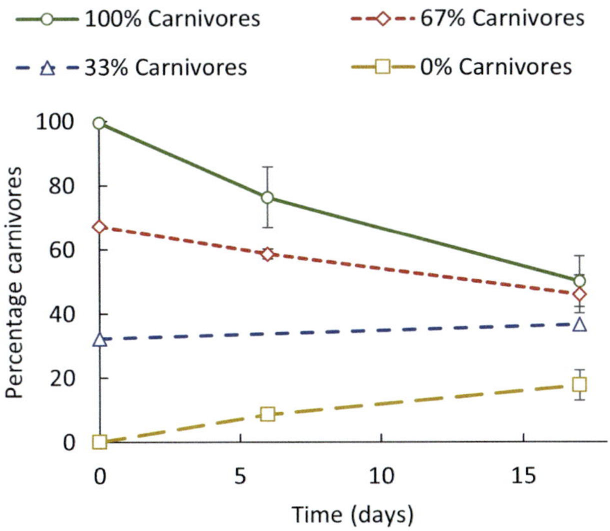

Figure 4.4c Change in percentage of carnivores over 17 days in pools with manipulated starting proportions of carnivores. Points show mean percentages of carnivores ± 1 SE. (Adapted from Pfennig, 1992. Reproduced with permission, copyright 1992 The Society for the Study of Evolution.)

Question 4.4.4: Pfennig checked the numbers of each morph at the end of the experiment and demonstrated that the results weren't caused by differential mortality of either morph. Why did so many carnivorous morphs in the 100% starting carnivore pools change to being omnivorous? (No more than 30 words are needed.)

Exercise 4.5: Lack clutch size

Extensive research on bird breeding around the world has demonstrated a trade-off between the number of offspring parents can produce and the amount of investment needed to rear those offspring to adulthood. This is known as the Lack clutch size hypothesis (named after David Lack, a world-famous ornithologist who proposed this trade-off in 1947). Experimental work which added or removed eggs or young birds from nests has produced the data shown in Table 4.5.

Table 4.5 Numbers of young birds surviving to recruitment in relation to addition or removal of eggs

Eggs added or removed	Young birds recruited to adult population	SE
–4	0.29	0.098
–3	0.55	0.079
0	0.85	0.054
3	0.64	0.128

Source: Pettifor et al., 2001.
Note: Eggs were taken from (negative) or added to (positive) nests.

Question 4.5.1: Which of these statements is the most appropriate explanation of these data?

A. If eggs were added to nests, more birds were recruited to the adult population.
B. Birds tend to lay the maximum number of eggs that they can rear to adulthood.
C. If eggs were removed from nests, survivorship of remaining birds is increased.

Question 4.5.2: However, further research (e.g. Monaghan and Nager, 1997) suggests that many birds are able to raise more offspring than the supposed optimal number. What additional factors might Lack have overlooked? (No more than 40 words are needed.)

Exercise 4.6: Limiting factors in wood pigeons

In the 1960s, a government policy was introduced in the UK to reduce losses of food crops to wood pigeons (*Columba palumbus*). Despite thousands of wood pigeons being shot over several years, the annual population count at the end of the breeding season remained the same and crop damage also remained the same. The policy was abandoned, and populations and damage to crops remained at the same levels.

Question 4.6.1: Which three of the options are most likely to account for these findings?

A. Wood pigeons were being killed by a disease.
B. Other species moved into the gaps created when wood pigeons were culled, increasing interspecific competition and resulting in the same level of crop losses.

C. Before and after the culling programme, wood pigeons experienced high intraspecific competition for food which reduced survival.
D. During the cull, wood pigeons experienced reduced intraspecific competition for food.
E. Culling resulted in unexploited food supplies which encouraged movement of wood pigeons from the surrounding area to take advantage of this.
F. Numbers of wood pigeons reduced so much that they were unable to increase again after the end of the cull.

Exercise 4.7: Pecking order in oystercatchers

In a coastal area in the Netherlands, researchers studied the breeding success of oystercatchers (*Haematopus ostralegus*). Although the young of most waders are *precocial* (they are able to find their own food from a very young age), oystercatchers are *semi-precocial* and need to bring food to their young. This allows them to nest farther away from feeding grounds in some situations.

In this study, some oystercatchers nested in territories on salt marsh (adjacent to feeding grounds) and others had territories farther inland, away from the feeding grounds. Salt marsh nesting birds had feeding territories adjacent to their nesting territories higher on the mud flat which were exposed by the tide for longer but had lower quality food. They raised 0.74 ± 0.184 young to fledgling stage. Inland nesters had feeding territories farther down the mudflats which were exposed by the tide for shorter periods of time but contained higher quality food; they raised 0.182 ± 0.039 young to fledging.

Figure 4.7 gives results for each stage of chick development.

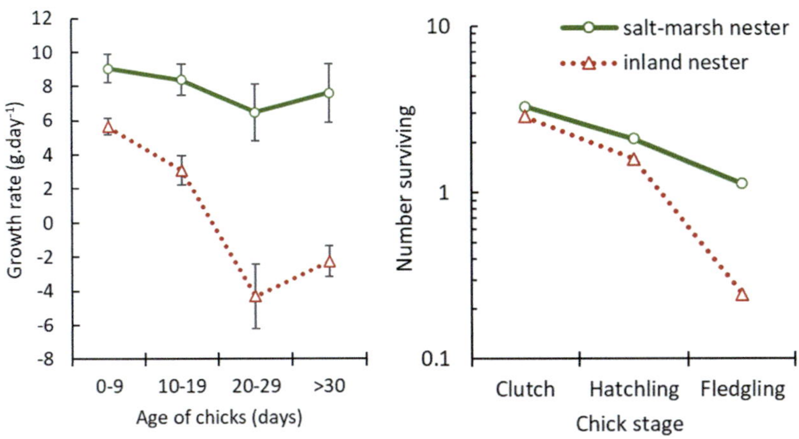

Figure 4.7 (Left) Growth rate of oystercatcher chicks over development period. (Right) Survival of young birds from eggs (clutch) through to fledging as adult birds (note log scale on the y-axis). (Adapted from Ens et al., 1992. With permission of John Wiley & Sons – Books.)

Question 4.7.1: Choose the five answers which best explain the data shown in this exercise.

A. Salt marsh birds raise more offspring than birds nesting inland.
B. Salt marsh birds raise fewer offspring than birds nesting inland.
C. Inland birds have to fly farther, which uses more energy.
D. Salt marsh territories have a greater value and are probably occupied by more-experienced birds.
E. Salt marsh birds have to spend more time defending their territories.
F. More-experienced birds may be better at providing food for their chicks.
G. There are higher levels of predation inland.
H. The data suggest that there are more places for young birds to hide on salt marshes and thus avoid being predated upon.
I. There is more food available on the salt marsh than on the inland fields so young birds are able to mature faster.

Table 4.7 Capture rates of food by birds nesting in different habitats

Capture rate of food at different stages of breeding	Salt marsh nesters			Inland nesters		
	Mean	SD	*n*	Mean	SD	*n*
Pre-laying	1.94	0.93	7	2.62	0.52	14
Incubation	2.85	1.06	32	3.32	1.20	21
Chick-feeding	2.53	0.58	17	3.23	1.42	38

Source: Ens et al., 1992. With permission of John Wiley & Sons – Books.
Note: *n* is the number of tides over which observations were made.

Question 4.7.2: The data in Table 4.7 seem to contradict the data in Figure 4.7. How could you explain what is happening? (No more than 60 words are needed.)

Exercise 4.8: Control of noisy miners

The noisy miner (*Manorina melanocephala*) is native to Australia and described as 'overabundant.' It has benefited from landscape changes due to farming activity and can reach high population levels in remnant open canopy eucalyptus forest. Its aggressive, competitive behaviour threatens less common bird species, and some conservationists advocate culling the species to reduce abundance. It has been suggested that noisy miners start to affect woodland bird communities when population densities exceed 0.8 noisy miners.ha^{-1}.

Researchers quantified the abundance of noisy miners for 10 months before and 12 months after 'patch culls,' covering a total area of 200 ha of woodland (average size of patch was 13 ha, ranging from 4 to 49 ha). The cull took 2 days and removed >500 birds (the highest population counted across all the sites during the pre-cull surveys) (Figure 4.8).

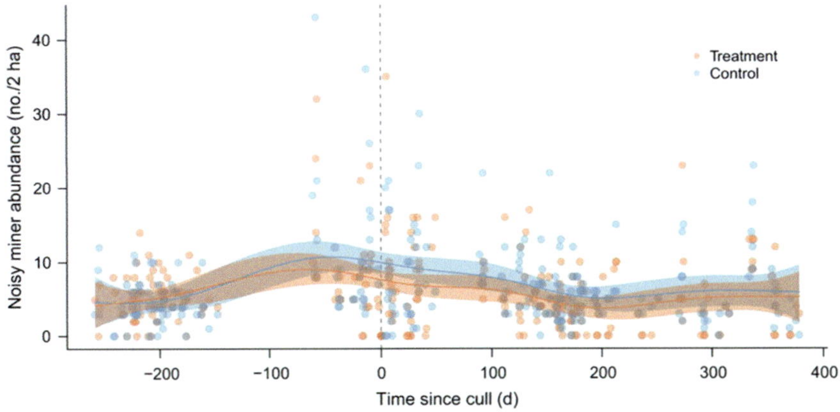

Figure 4.8 Abundance of noisy miners before and after a cull (treatment); no culling took place in control plots. Points represent single-site surveys; the fitted lines show 95% confidence intervals for areas where culls took place and control areas. (From Beggs et al., 2019. Reproduced with permission, copyright 2019 by the Ecological Society of America.)

Question 4.8.1: Using the graph in Figure 4.8, was the cull likely to have been successful in conserving woodland bird communities?

Question 4.8.2: Which of the following statements is the most likely explanation for the result of this experiment?

A. Noisy miners bred more successfully due to reduced intraspecific competition and raised more chicks to adulthood.
B. Adult birds quickly moved into vacant areas from suboptimal habitat.
C. There is natural variation in population size throughout the year.

Question 4.8.3: In no more than 40 words, how might land managers make use of these results to conserve woodland bird communities?

5

Interspecific competition

Exercise 5.1: Competition between pink muhly grass and native species in South Korea

Pink muhly grass (*Muhlenbergia capillaris*) is native to the US. While it is considered to be endangered in some parts of its native range, in other areas there is concern that it could become invasive and displace native flora. In South Korea, pink muhly grass has been used extensively in landscaping projects, but the Ministry of the Environment now requires sites to be monitored to assess its potential to become problematic.

Byun (2023), from Andong National University, South Korea, conducted controlled greenhouse experiments to investigate the impact of pink muhly grass on 17 native grassland plants. Native plants were grown in monoculture or in two-way combinations (i.e. pink muhly grass with one other species); percentage cover was recorded in both years and biomass was determined at the end of the experiment. There was no difference between the above-ground biomass of native plants when competing with pink muhly grass and growing in monoculture ($F_{1,98} = 1.17$; $P = 0.280$) (Figure 5.1).

DOI: 10.1201/9781003644095-5

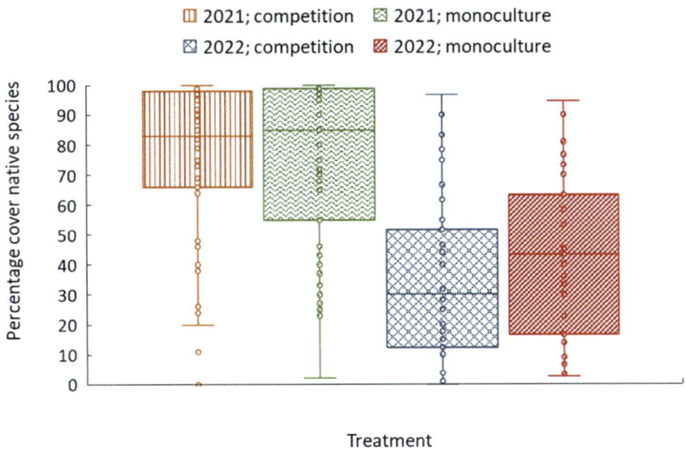

Figure 5.1 Box and whisker plot showing results from 2 years of data to assess the competitive impact of pink muhly grass on native species. (Adapted from Byun, 2023. Reproduced under the Creative Commons CC BY licence, copyright 2023 Byun.)

Question 5.1.1: Based on these results, should the South Korean Ministry of the Environment be concerned about the invasive potential of pink muhly grass?

When Byun compared results for each species in monoculture and with pink muhly grass, two species showed a significant decrease in interspecific competition compared with intraspecific competition: China pink (*Dianthus chinensis*) and Chinese fountaingrass (*Cenchrus alopecuroides*; formerly known as *Pennisetum alopecuroides*).

Question 5.1.2: What additional factors should be considered before making a definitive decision about the potential for pink muhly grass to become a problematic invasive species in South Korea? Consider experimental design issues (at least two issues) and wider environmental issues (at least three issues). (Aim for 100 words to give a complete answer.)

Exercise 5.2: Interactions between bluebirds

In the Western US, mountain bluebird (*Sialia currucoides*) and western bluebird (*S. mexicana*) have overlapping ranges and appear to compete for nest sites. Both species rely on forest fires to open up the understory and create meadows that they use for feeding. The standing dead wood created by fires provides cavities in which the birds nest. Such areas remain suitable for about 30 years until natural succession leads to meadows being replaced by scrub woodland. Natural nest sites are in very short supply, but both species have switched to using nest boxes provided by people in urban areas and farmland. This has provided opportunities to investigate species preferences for nest boxes located in transects at high density (75–100 m apart; N = 4) and those at low density

(150–200 m apart; N = 4). 'Interior' boxes were in the middle of a line of boxes and 'exterior' ones were at either end.

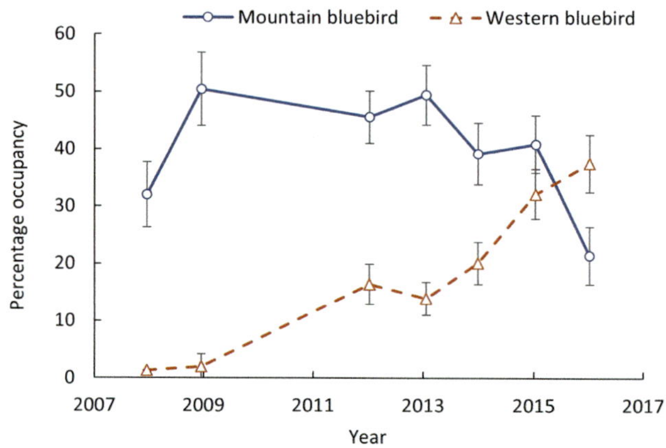

Figure 5.2a Percentage occupancy of nest boxes by mountain and western bluebirds from 2008 to 2016. Points show mean ± 1 SE. (Adapted from Duckworth et al., 2017. Reproduced under the Creative Commons CC BY licence, copyright 2017 Duckworth et al.)

Question 5.2.1: Studies suggest that there is a trade-off between colonising ability and competitiveness. Using Figure 5.2a, which species appears to be the stronger competitor: mountain bluebird or western bluebird?

Question 5.2.2: Which is the better coloniser: mountain bluebird or western bluebird?

Despite differences in the strength of their interactions, mountain and western bluebirds are able to coexist for periods of over 10 years. Mountain bluebirds are larger and need bigger territories; the study revealed they had a distinct preference for nesting sites (Figure 5.2b).

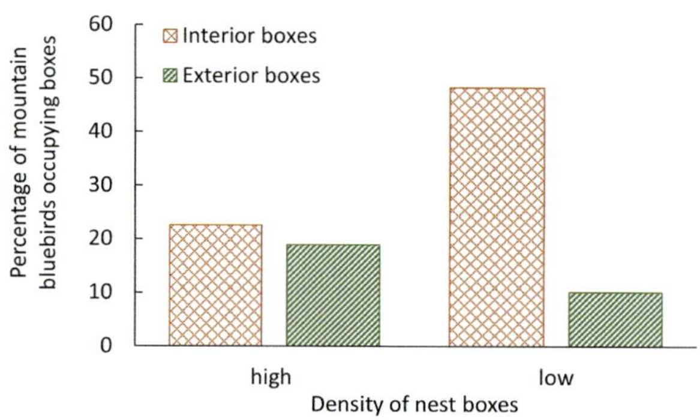

Figure 5.2b Preferences of mountain bluebirds for position and density of nest boxes. (Adapted from Duckworth et al. 2017. Reproduced under the Creative Commons CC BY licence, copyright 2017 Duckworth et al.)

Question 5.2.3: The researchers had expected the opposite effect in relation to the position of boxes and no difference due to the density of boxes. In contrast, western bluebirds showed no pattern in relation to the density or position of boxes. Which of the following statements is most likely to explain what is happening here? (More than one answer is correct.)

A. Competition isn't the only factor affecting nest site preference.
B. Competition occurs at a smaller scale than 150 m, so there are fewer interactions between birds nesting in adjacent boxes.
C. There is a trade-off between reduced competition at the ends of the transects and higher predation where there are fewer birds to warn of potential threats.
D. The layout of the boxes and number of replicates may need amending to be able to show patterns more distinctively.

Question 5.2.4: Mountain bluebirds fledge more young than western bluebirds. How might this be linked with competitive behaviour? (No more than 20 words needed.)

Exercise 5.3: Competition between water fleas

Water fleas (*Daphnia* spp.) are planktonic aquatic crustaceans, up to 6 mm long. In the absence of predation, competition is considered to be a key interaction structuring populations. Researchers cultured two species of water flea: *Daphnia magna*, a large-bodied species; and *D. longispina*, a small-bodied species. They started the 400 ml cultures containing both 5x *D. magna* and 20x *D. longispina* (different starting sizes of the populations ensured a similar biomass of each species). Two sets of cultures were set up using algae as a food source: one had a high P concentration and the other low P concentration.

Question 5.3.1: The results were analysed using a repeated measures ANOVA (this is used for datasets where we expect a change over time, but the focus is on differences between treatments). There was a significant interaction term between species and P concentration ($F_1 = 57.3$, $P < 0.0001$). What does this result mean?

A. Populations of *D. magna* grew faster than those of *D. longispina.*
B. Populations of *D. longispina* grew faster than those of *D. magna.*
C. The species responded differently to the high and low P treatments.
D. *D. magna* was the stronger competitor in the low P treatment.
E. *D. longispina* was the stronger competitor in the high P treatment.
E. *D. magna* was the weaker competitor in the low P treatment.
F. *D. longispina* was the weaker competitor in the high P treatment.

Figure 5.3 shows the detailed results from the experiment by Feniova et al. (2025).

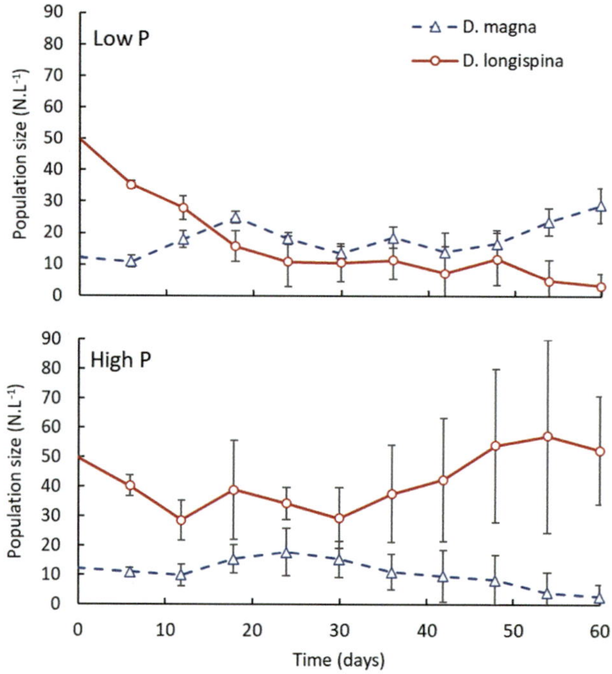

Figure 5.3 Time course of competition experiment between *D. magna* and *D. longispina* over 60 days at low and high P. Points show mean ± 1 SD. (Adapted from Feniova et al., 2025. Reproduced under the Creative Commons CC BY licence, copyright 2025 Feniova et al.)

Question 5.3.2: The researchers hypothesised that *D. magna* would be the stronger competitor in the high P treatment as it has higher growth rates and *D. longispina* would be stronger in the low P treatment as it is better able to tolerate low food supplies. Were they correct?

Question 5.3.3: In eutrophic waters in natural systems *D. magna* is often the dominant species. It has also been observed that this species is a better coloniser than *D. longispina*. Explain how the results from this experiment fit with this extra information? (No more than 100 words are needed.)

Question 5.3.4: The largest loss of organismal P occurs when *Daphnia* moult their skins. Why might this have a greater impact on small *Daphnia* species? (No more than 25 words are needed.)

Exercise 5.4: Interactions between forest ants

Mound-building red wood ants are a dominant group of ants in boreal forests in Finland. Colonies of the Scottish wood ant (*Formica aquilonia*), a more northern species, and the European red wood ant (*F. polyctena*), a more southerly species, both construct a number of nests within their territory and thus require larger forest patches than species in which each colony has only one nest. Researchers investigated interactions between these two species in an area where they overlap. They sampled forest patches to determine if they were occupied by one or both species of ant or were unoccupied. The data were analysed using a type of chi-square test (Figure 5.4).

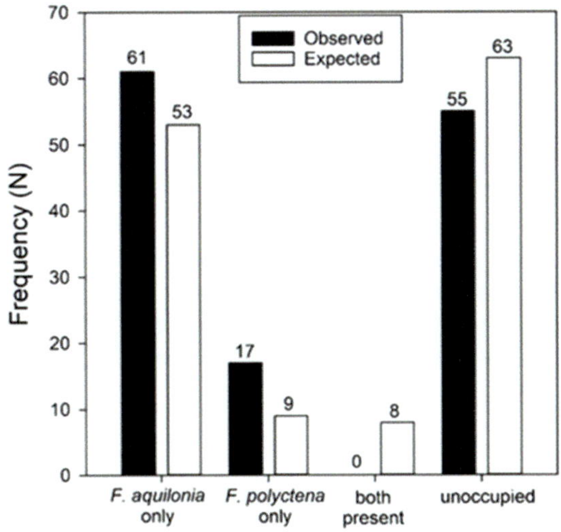

Figure 5.4 Observed and expected results of *F. aquilonia* and *F. polyctena* ants when found together or separately in forest patches. (From Sorvari et al., 2024. Reproduced under the Creative Commons CC BY licence, copyright 2024 Sorvari et al.)

Here is the chi-square test result: $N = 133$, $df = 3$, $G^2 = 14.70$, $p = 0.0021$.

Question 5.4.1: Using the data given, which of the following statements are most likely to be true? (Two answers are correct.)

A. *F. aquilonia* and *F. polyctena* have no influence on the distribution of the other species.
B. *F. aquilonia* is more common than *F. polycten*a in this forest.
C. The large number of unoccupied forest patches suggests that these species of ants have driven each other to extinction in these areas.
D. These ant species do not coexist.

Question 5.4.2: What term could we use to describe the interaction between these ants?

A. Niche partitioning
B. Character displacement
C. Competitive release
D. Competitive exclusion

Question 5.4.3: Ecological theory suggests that differences in the dispersal ability of species can influence their ability to coexist in fragmented habitats. In no more than 70 words, explain how this might be relevant to these ant species in Finland. What name do we give to these less competitive but better dispersing species?

Exercise 5.5: Coexistence in seabirds

Seabird populations are generally regulated by bottom-up processes (i.e. food supply rather than predation). As their food sources tend to occur patchily in the environment, these birds often live in large colonies. The downside of this is that intra- and interspecific competition for food can be very intense. One option to overcome this issue is to forage in different places (Figure 5.5a).

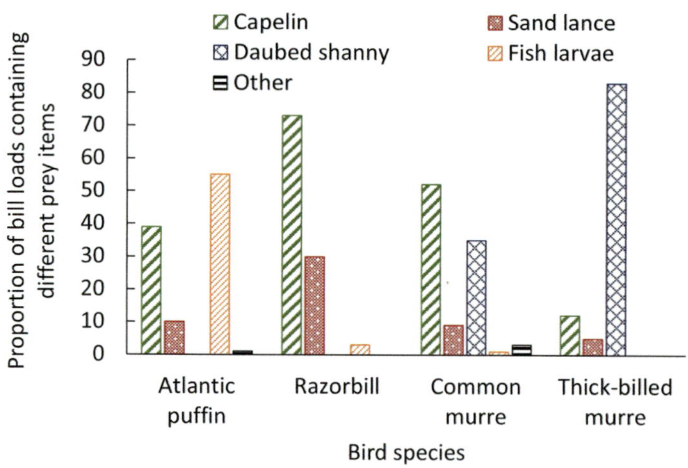

Figure 5.5a Food types in the beaks of auks feeding chicks. Bars show means. (Adapted from Pratte et al., 2017. Reproduced with permission of the Marine Ecology – Progress Series.)

Stable isotopes of C and N in the blood of adult birds give an indication of the prey on which the birds had been feeding (Figure 5.5b).

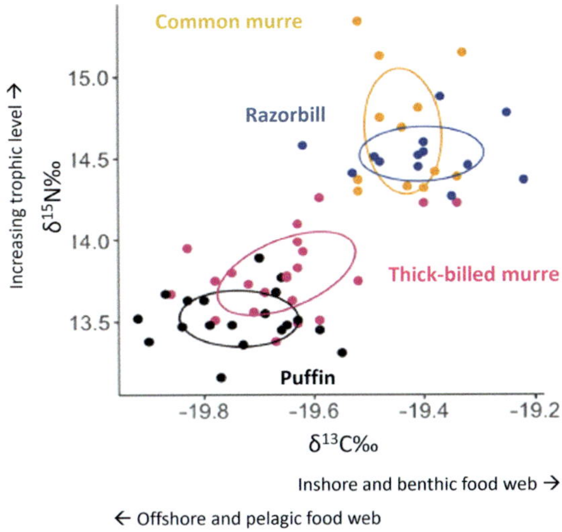

Figure 5.5b $\delta^{15}N‰$ and $\delta^{13}C‰$ from blood samples of adult birds. The ellipses have been calculated to show isotopic niche width for each species and degree of overlap. (Adapted from Pratte et al., 2017. Reproduced with permission of the Marine Ecology – Progress Series.)

Question 5.5.1: Which of the following terms is the most appropriate description of these results?

A. Niche partitioning
B. Character displacement
C. Competitive exclusion
D. Cooperation
E. Interference competition
F. Exploitation competition
G. Apparent competition

Question 5.5.2: Explain in no more than 50 words how common murre and razorbill can coexist.

Exercise 5.6: Seasonal changes in coexistence in ducks

A study of seven species of dabbling ducks in North America, all in the genus *Anas*, made comparisons between pairs of species and the degree of overlap in food they ate and habitat in which they spent their time. Figure 5.6 shows the results of this work.

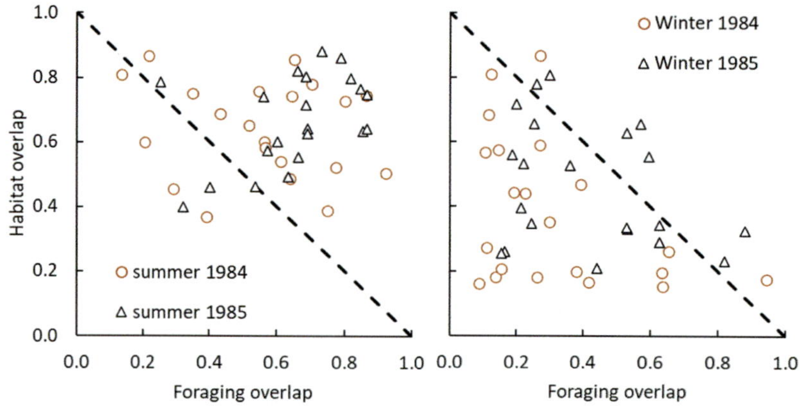

Figure 5.6 Overlap between habitat use and foraging in summer (left) and winter (right); each point shows the relationship between a pair of species: 0 shows there is no overlap and 1 represents complete overlap. (Adapted from DuBowy, 1988. Reproduced with permission, copyright 1988 by the Ecological Society of America.)

Question 5.6.1: The dashed line running diagonally across the graphs isn't a line fitted to the data. It indicates an important feature of the graphs. What does it show? Choose the most appropriate answer from the following options.

A. It indicates a strong negative association between food choice and habitat choice.
B. It indicates a strong positive association between food choice and habitat choice.
C. Below the line, there is less overlap of food and habitat use.
D. Above the line, there is less overlap of food and habitat use.
E. It shows that competition is occurring.
F. It shows that competition is not occurring.

Question 5.6.2: In no more than 40 words, explain how the relationship between the ducks changes according to the season and suggest why this might occur.

Exercise 5.7: Gause, Paramecia and Lotka–Volterra

A PhD student has decided to replicate some of Gause's experiments using two *Paramecium* species (shown as sp1 and sp2) to see if she can apply Lotka–Volterra models to her results. She set up four trials of culturing the two species together using different starting population sizes and measured the population size of each species over 21 days. Apart from different numbers of *Paramecium* to start each trial, all other experimental conditions were constant. Her results are shown in Figure 5.7.

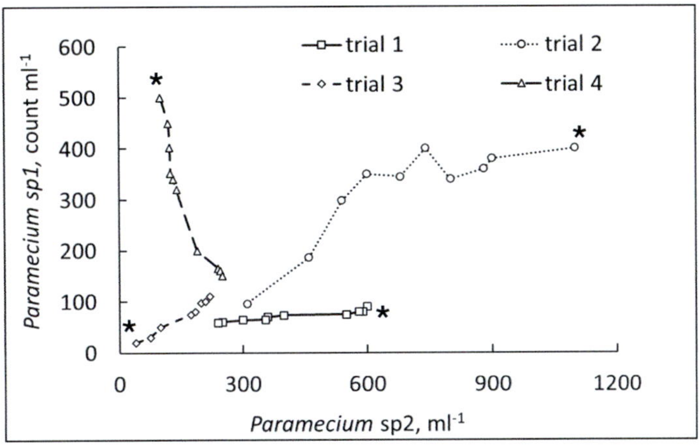

Figure 5.7 Results of trials with two species of *Paramecium* over 21 days using different starting numbers of populations. Values are numbers of individuals ml^{-1}. The starting population sizes of each trial are marked with an asterisk.

Question 5.7.1: Which outcome from Lotka–Volterra competition models is the most likely one to explain these results?

A. Spl drives sp2 to extinction.
B. Sp2 drives spl to extinction.
C. Spl and sp2 have a long-term stable coexistence.
D. Spl and sp2 coexist in the short term only.

Exercise 5.8: Interactions between mink and otters

Eurasian otters (*Lutra lutra*) disappeared from much of the UK around 100 years ago due to hunting and pesticide poisoning from DDT. During the absence of otters, American mink (*Neogale vison*) escaped from various fur farms and established themselves across much of the UK. Otters have now recolonised much of their historic range. Otters are largely aquatic, whereas mink are a more generalist species. Research suggests that mink were able to exploit a vacant niche but otters will now largely displace them.

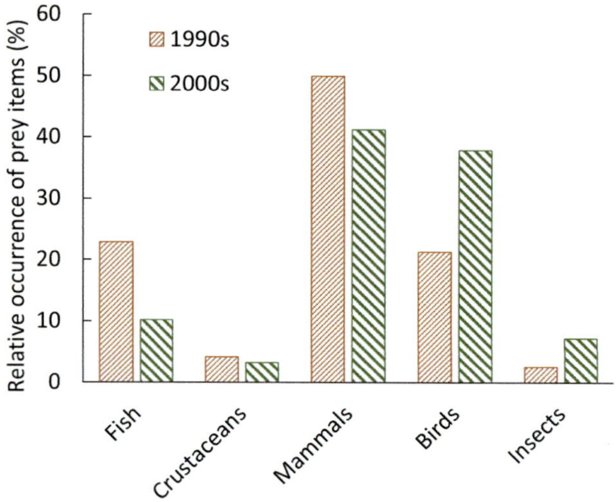

Figure 5.8a Changes in mink diet, determined from analysis of scat. (Adapted from Harrington et al., 2009. Reproduced with permission, copyright 2009 by the Ecological Society of America.)

Question 5.8.1: Otters weigh 6–10 kg and mink weigh around 1 kg. What might explain the changes in diet shown in Figure 5.8a? (Only 15 words are needed.)

Question 5.8.2: What term could be used to describe the pattern of prey selection between mink and otters? (Use only two words.)

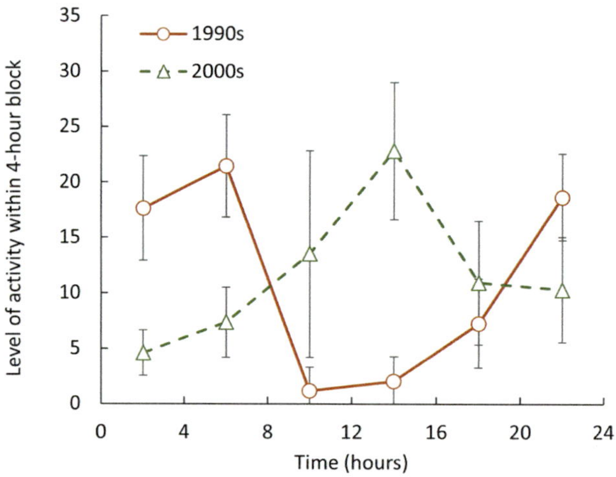

Figure 5.8b Changes in mink activity, recorded from animals with radio collars. Points are mean percentage active radio locations ± 1 SE. (Adapted from Harrington et al., 2009. Reproduced with permission, copyright 2009 by the Ecological Society of America.)

Question 5.8.3: Using Figure 5.8b, what is the most likely reason that mink have changed their foraging patterns between the two time periods?

A. Otters now eat the fish which mink would have caught, so mink have had to change their hunting behaviour.
B. Mink have been observed to have worse 'body condition' in the 2000s (measured as mass in relation to skeletal body size); this means they will spend less time hunting in water, which requires considerable energy to keep warm.
C. To avoid interference competition with otters.
D. It's easier to hunt birds during the day.

Question 5.8.4: What term could be used to describe the pattern of activity between mink and otters? (Use only three words.)

Question 5.8.5: The main mammal prey of mink was rabbits and wood mice, with some vole, rat and shrew species. The types of food item didn't change between the 1990s and 2000s. Why might the *proportion* of mammals in mink diets have reduced over the study period?

A. A shift in areas where mink forage away from rivers to more terrestrial areas has reduced opportunities for preying on mammals.
B. There has been a strong shift towards preying on birds, which are a better food source than mammals.
C. Mink preyed substantially on water voles but there are now none left in these areas, so mink have to find alternative food sources.
D. Change to more diurnal activity may reduce opportunities for preying on largely nocturnal mammal species.

Question 5.8.6: Do these data suggest that otters will displace mink, as conservationists have been hoping?

6

Mutualism and parasitism

Exercise 6.1: Elephants, ants and scale insects

The whistling thorn (*Vachellia drepanolobium*; previously known as *Acacia drepanolobium*) is a tree found in East Africa and is a key constituent of woodlands. In common with other *Vachellia* species, it has a relationship with ants, in this case, *Crematogaster mimosae*. The tree provides the ant with nectar, and some of the thorns are swollen and provide shelter for ants. The ants protect trees from grazing by herbivores, which can cause considerable damage to thorn trees. *Crematogaster mimosae* don't get all their food from nectar provided by the tree; they also tend scale insects (sap-sucking insects found on leaves) for carbohydrate-rich honeydew (Figure 6.1).

DOI: 10.1201/9781003644095-6

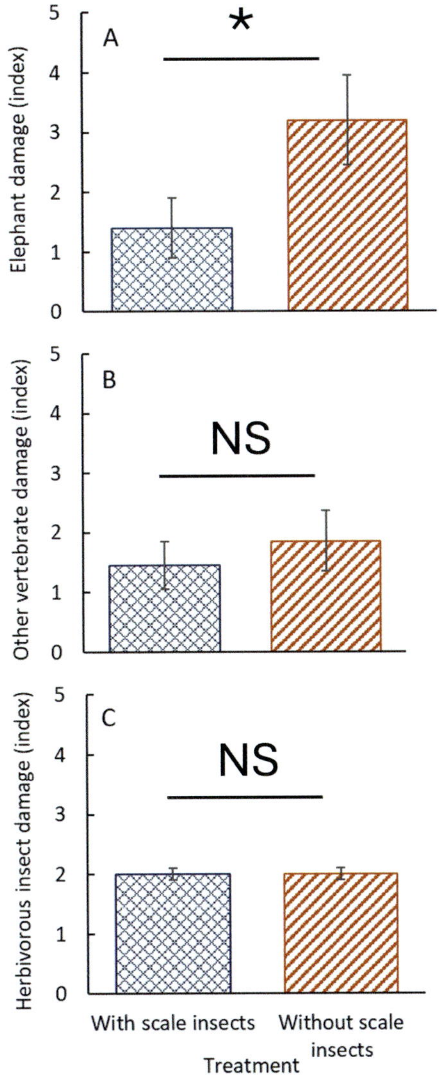

Figure 6.1 Damage to whistling thorn trees from elephants (A), other vertebrate browsers (B) and herbivorous insects (C). Damage is shown using means of indices (scale of 1–4) ± 1 SE over 10 months. (Adapted from Prior and Palmer, 2018. With permission of John Wiley & Sons – Journals.)

Question 6.1.1: What is the most appropriate description of the relationship between *Crematogaster* ants and whistling thorn trees? (Choose all that could apply.)

A. Parasitism (one gains, one suffers)
B. Mutualism (both gain)

C. Amensalism (one suffers, the other doesn't gain anything)
D. Commensalism (one gains, the other doesn't suffer)
E. Neutral (neither gain)

Question 6.1.2: What is the most appropriate description of the relationship between *Crematogaster* ants and scale insects? Choose all that could apply.

A. Parasitism (one gains, one suffers)
B. Mutualism (both gain)
C. Amensalism (one suffers, the other doesn't gain anything)
D. Commensalism (one gains, the other doesn't suffer)
E. Neutral (neither gain)

Question 6.1.3: What is the most appropriate description of the relationship between whistling thorn trees and scale insects? (Choose all that could apply.)

A. Parasitism (one gains, one suffers)
B. Mutualism (both gain)
C. Amensalism (one suffers, the other doesn't gain anything)
D. Commensalism (one gains, the other doesn't suffer)
E. Neutral (neither gain)

Question 6.1.4: What are the broader impacts of scale insects on this woodland ecosystem? (No more than 30 words are needed.)

Question 6.1.5: What is the most likely mechanism by which scale insects deter elephant grazing?

A. Ants are more active as they gain additional food from tending scale insects.
B. Scale insects have a bad taste and deter grazers.
C. Ants feed on scale insects so they are more active and aggressive towards grazers.
D. Scale insects induce the growth of protective thorns by the tree.

Question 6.1.6: Why does the presence or absence of scale insects make no difference to the damage to whistling thorn trees from other browsers and invertebrate grazers? (No more than 50 words are needed.)

Exercise 6.2: Grazing by marsh periwinkles

Marsh periwinkle snails (*Littoraria irrorata*) graze on saltmarsh cordgrass (*Sporobolus alterniflora*; formerly known as *Spartina alterniflora*) on the Southeast coast of the US. Close observation has shown that the periwinkles do not consume live tissue directly. Instead, they create and maintain wounds along the leaf surface and feed on ascomycete fungi that colonise the dead material around the wounds. Researchers have also observed that periwinkles preferentially defecate on grazed sections of leaf.

The top graph of Figure 6.2 shows results from an experiment where leaves from each treatment (ungrazed, grazed or simulated grazing) were fed to juvenile periwinkles, and their growth was measured. The bottom graph shows fungal biomass from leaves which were ungrazed, leaves exposed to approximately 220 periwinkles.m^{-2} or leaves which had been scraped with a razor to simulate grazing wounds.

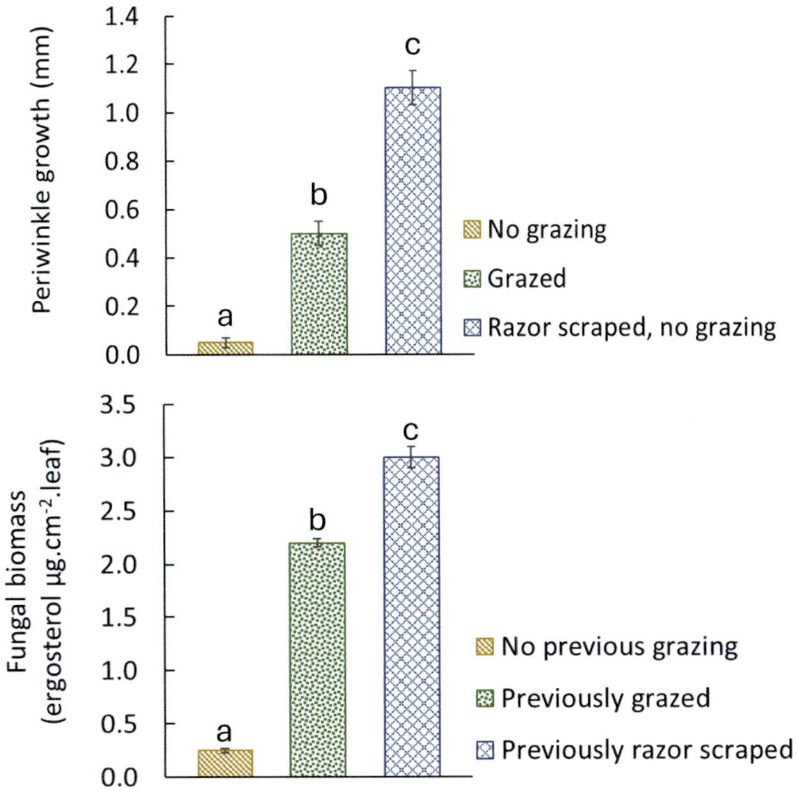

Figure 6.2 Marsh periwinkle growth (top) and fungal biomass (bottom) in relation to grazing level or physical scraping damage. Bars show mean ± 1 SE. Bars sharing a letter are not significantly different from each other. (Adapted from Silliman and Newall, 2003. Reproduced with permission, copyright 2003 National Academy of Sciences, U.S.A.)

Question 6.2.1: Which of the following statements is the most appropriate description of these results?

A. Marsh periwinkle snails preferentially eat saltmarsh cordgrass.
B. Leaf damage allows colonisation by fungi.
C. Periwinkles' growth is greater when they feed on fungi rather than saltmarsh cordgrass leaves.
D. The presence of periwinkle faeces around grazing wounds deters periwinkles from grazing.

Question 6.2.2: How could you describe this relationship between periwinkles and fungi? Can you think of any similar examples from other parts of the world? (No more than 50 words are needed.)

Question 6.2.3: Why might there be more fungal growth on the leaves in the simulated grazing treatment? Hint: remember that there are no periwinkles present in this treatment. (No more than 25 words are needed.)

Question 6.2.4: Having figured out the answer to the previous question, why is there more periwinkle growth on the leaves that received simulated grazing rather than actual periwinkle grazing? (Only 25 words are needed.)

Exercise 6.3: Marsh periwinkles, grazing and parasites

Salt marshes are important coastal habitats which are regularly inundated at high tides. The vegetation is dominated by grasses, herbs or shrubs which play a key role in trapping sediment, thus protecting land from storms. They are also important stores of C.

As set out in the previous question, on salt marshes in the south of the US, the marsh periwinkle (*Littoraria irrorata*) is an important grazer of saltmarsh cordgrass (*Sporobolus alterniflora*; formerly known as *Spartina alterniflora*). Heavy grazing by the periwinkle during drought conditions can lead to large areas of die-off of salt marsh plants. Marsh periwinkles are commonly infected by the parasitic trematode worm *Parorchis acanthus*.

Figure 6.3 and Table 6.3 show results from an experiment to assess damage to saltmarsh cordgrass leaves. Salt was added to half of the experimental treatment plots to simulate the effects of drought.

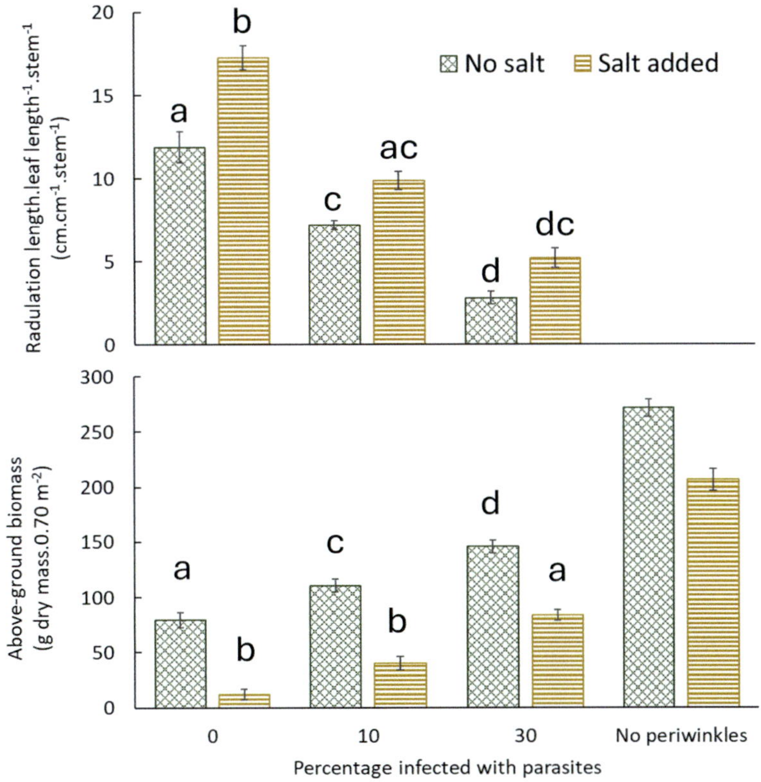

Figure 6.3 Effect of parasite infection. (Top) Radulation length.leaf length^{-1}.stem^{-1} is a measure of grazing damage by periwinkles on the saltmarsh cordgrass. (Bottom) Above-ground biomass of saltmarsh cordgrass plants with differing levels of parasitic infection. Bars show mean $\pm$ 1 SE. Letters above bars show results of Tukey post hoc tests; bars sharing the same letter are not significantly different from each other. Bars showing above-ground biomass where no periwinkles were present were not included in the analysis and are shown for comparison only. (Adapted from Morton and Silliman, 2019. Reproduced with permission, copyright 2019 by the Ecological Society of America.)

Table 6.3 Results of statistical tests with leaf damage and above-ground biomass as response variables and parasite prevalence and salt addition (simulated drought) as treatments

	Leaf damage				Above-ground biomass			
	MS	F	df	P	MS	F	df	P
Parasite prevalence (PP)	880.83	175.25	1, 44	<0.0001	37158.00	118.36	2, 42	<0.0001
Salt addition (SA)	132.10	26.28	1, 44	<0.0001	54002.00	172.02	1, 42	<0.0001
PP × SA	23.15	4.60	1, 44	0.037	98.00	0.31	2, 42	0.5786
Error	221.15				314.00			

Source: Adapted from Morton and Silliman, 2019. Reproduced with permission, copyright 2019 by the Ecological Society of America.

Question 6.3.1: Based on these results, which one of the following statements is correct?

A. Simulated drought reduces the effect of parasitic infection by *Parorchis* on saltmarsh cordgrass.
B. Above-ground biomass increases during simulated drought.
C. Periwinkles infected with parasitic *Parorchis* worms eat less than uninfected periwinkles.

Question 6.3.2: The analysis of the leaf damage data showed a significant interaction between parasite prevalence and salt addition. Use the graph in Figure 6.3 to explain what this means. (No more than 60 words are needed.)

Question 6.3.3: The graph showing above-ground biomass (Figure 6.3) and the corresponding section of the statistics table (Table 6.3) show that the interaction between parasite prevalence and salt addition was not significant. Choose the most appropriate explanation of this from the following options.

A. There is no significant difference between the salt addition treatment for 0% and 10% infection with *Parorchis*, nor between salt addition at 30% infection with *Parorchis* and no salt added when there is 0% infection with the parasite.
B. There is a statistically significant difference between each pairwise comparison of response to drought at each level of parasite prevalence.
C. Above-ground biomass increases as parasite prevalence increases.
D. All the pairwise comparisons show the same pattern of response to addition of salt at each level of parasite prevalence.

Question 6.3.4: Parasites sometimes manipulate the behaviour of their hosts to increase the likelihood that the parasite will be spread to new hosts. In this example, infected periwinkles tend not to climb up the grass stems but stay at the base. How might this aid the spread of the parasite? (No more than 40 words are needed.)

Exercise 6.4: Anemone symbiosis

Researchers investigated the mutualistic relationship between the sea anemone (*Exaiptasia pallida*) and the dinoflagellate (*Breviolum minutum*). This relationship appears to be very similar to that between corals and their mutualistic dinoflagellate partner, but anemones are more amenable to culturing and manipulation in the lab. They set up cultures of anemones and cold-shocked them until all the dinoflagellates had been lost. Using this set of anemones, they set up four treatments: two cultures were maintained without any dinoflagellates ('aposymbiotic'), the other two were inoculated with

Breviolum ('colonised'). One from each symbiosis type was fed regularly and the other was starved, giving a fully factorial experimental design (Figure 6.4).

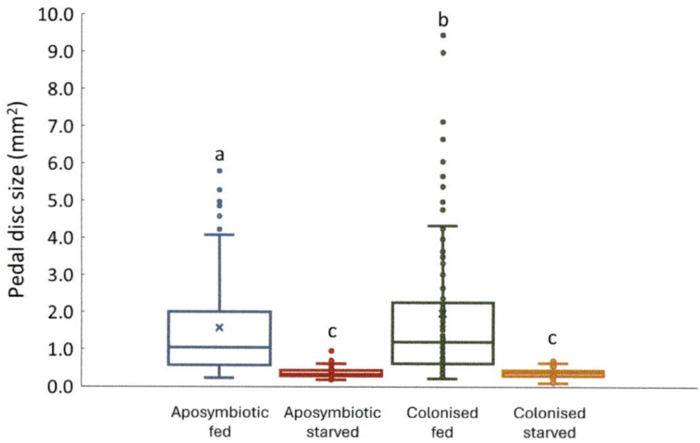

Figure 6.4 Box and whisker plot showing data from weekly sampling points over 2 months. The y-axis shows the size of pedal discs (the flat plate on the opposite side from the tentacles which anemones use to attach to their substrate) of anemones colonised or not by *Breviolum* dinoflagellates and subsequently fed or starved. Lowercase letters above the bars show results from Tukey post hoc tests; bars sharing a letter are not significantly different. (Adapted from Peng et al., 2020. Reproduced under the Creative Commons Attribution Licence, copyright 2020 Peng et al.)

Question 6.4.1: Select the correct interpretations of these data from the following list.

A. There is an obligate mutualistic relationship between *Exaiptasia* anemones and *Breviolum*.
B. *Breviolum* are parasitic rather than mutualistic.
C. There is no value to *Exaiptasia* anemones in having the *Breviolum* symbiont as shown by its ability to grow very well alone as long as there is sufficient food available.
D. The relationship between *Exaiptasia* anemones and *Breviolum* changes according to whether *Exaiptasia* are fed or not.

Question 6.4.2: After 6 months, the anemones in the cultures which had been starved had either died or formed a tissue ball without any identifiable structure. In contrast, dinoflagellates were still present, alive and dividing in the tissue. What might explain this finding? (No more than 40 words needed.)

Question 6.4.3: It's suggested that coral bleaching occurs when the cnidarian partner expels the dinoflagellate because the latter stops supplying the cnidarian with C compounds from photosynthesis. How could you develop the experiment described here with *Exaiptasia* anemones to investigate whether similar interactions occur? (No more than 40 words are needed.)

Exercise 6.5: Invasive crayfish in North America

The spiny stream crayfish *Faxonius cristavarius* is native to West Virginia but was introduced to Mountain Lake, Virginia, in the 1930s. Since then, it has spread to nearby river catchments. It can displace the native crayfish found in this area—several species of the genus *Cambarus*. The native species act as hosts to crayfish worms (branchiobellida); the worms are obligate symbionts, cannot reproduce without the crayfish and can only spread by direct contact between hosts. The carapaces of older crayfish build up layers of biofilm (protozoa and metazoa), and the crayfish worm feeds on this, performing a 'cleaning service' which can reduce mortality and increase crayfish growth rates. Younger crayfish moult more frequently and thus have cleaner carapaces so colonising worms feed on the host's tissue. In response, young hosts frequently groom themselves to remove the worms. In their *native* range, the spiny stream crayfish has similar symbiotic relationships with branchiobdellidan worms, but in their *invasive* range, old and young crayfish regularly groom themselves to remove the worms. A worm that is removed by a crayfish is highly unlikely to find a new host individual.

Bell and colleagues (2024) investigated the impacts of the invasive spiny stream crayfish on crayfish worm communities from a range of locations in the Mountain Lake region of Virginia. Figure 6.5 shows some of their results.

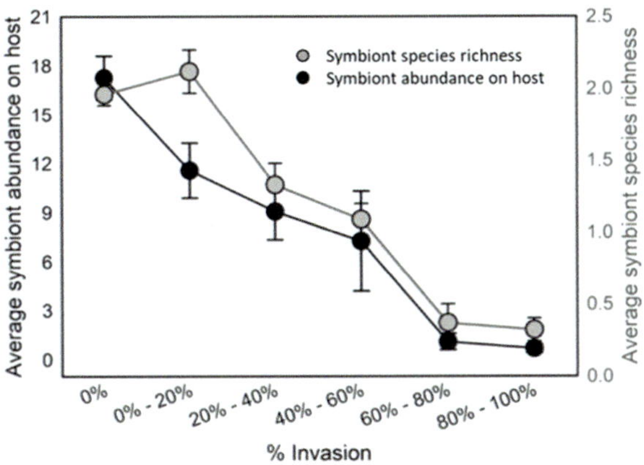

Figure 6.5 Symbiont abundance on native host crayfish (left-hand axis, black points); average species richness of symbiont community (right-hand axis, grey points) in relation to percentage of invasive crayfish. Data points are averages ± 1 SE. (From Bell et al., 2024. Reproduced under the Creative Commons CC BY licence, copyright 2024 Bell et al.)

Question 6.5.1: Which of the following statements gives the most appropriate overview of these data?

A. As the proportion of invasive crayfish increases, the proportion of native crayfish decreases.
B. As the proportion of invasive crayfish increases, invasive crayfish act as host to fewer species of symbiont and to fewer individual symbionts.
C. As the proportion of invasive crayfish decreases, symbiont abundance and diversity decreases.
D. As the proportion of invasive crayfish increases, symbiont abundance and diversity on host species decreases.

Question 6.5.2: Branchiobdellidan worms are not good at discriminating between good hosts (native crayfish) and poor ones (invasive crayfish). In addition, the invasive crayfish are more active than the native crayfish, which tend to hide under rocks. In no more than 70 words, set out how these two factors could reduce further the populations of native crayfish and their symbionts.

Question 6.5.3: What are the long-term implications of the arrival of this invasive species of crayfish, both for the native and invasive crayfish and the branchiobdellidan worms? (Seventy words will suffice as a complete answer.)

Exercise 6.6: Yucca and moths

Yucca plants (*Yucca filamentosa*) are pollinated by yucca moths (*Tegeticula yuccasella*). Inflorescences contain 150–500 individual flowers. Newly hatched, female moths collect pollen, lay their eggs in the ovary of the flower and then travel to the stigma and pollinate the flower. The moth larva feeds on the seed that develops in the ovary where the egg was laid. The yucca can selectively abort flowers (Figure 6.6).

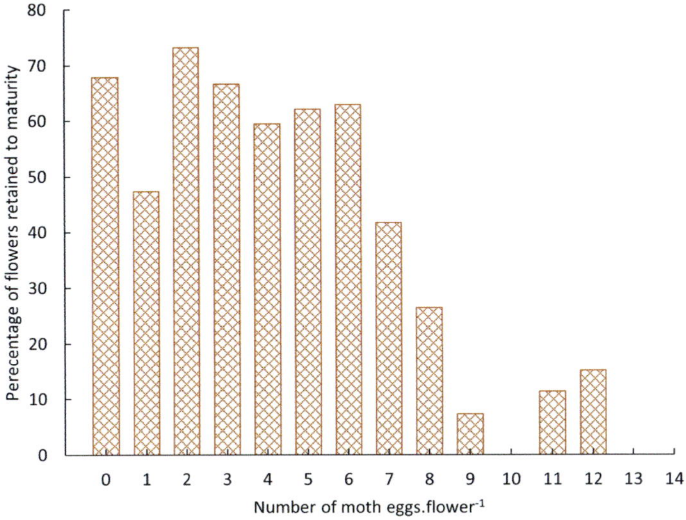

Figure 6.6 Percentage of yucca flowers retained to maturity in relation to the number of yucca moth eggs in each flower. Where flowers were aborted, the plant did this before moth eggs hatched. Bars show means. (Adapted from Pellmyer and Huth, 1994. Reproduced with permission, copyright 1994 Springer Nature Limited.)

Question 6.6.1: What is the best description of the interaction between moths and yucca plants when there are one to six eggs in an individual ovary?

Question 6.6.2: What is the best description of the interaction between moths and yucca plants when there are >13 eggs in an individual ovary?

Question 6.6.3: Why don't yucca plants retain 100% (or at least the highest proportion) of flowers with no or only one moth egg?

A. The plant has no mechanism for detecting numbers of moth eggs laid in flowers.
B. The flowers were probably not fully pollinated.
C. This suggests that resources are limited and the plant cannot afford to maintain all of the flowers.
D. It is more advantageous to the plant to retain flowers with greater numbers of moth eggs as this will lead to greater numbers of moths and thus more pollination in the next generation.

Question 6.6.4: Why does the yucca abort flowers with more than 12 moth eggs in them? (No more than 40 words are needed.)

Exercise 6.7: Impact of a parasitic plant on salt marshes

Marsh dodder (*Cuscuta salina*) is an annual, holoparasitic plant (i.e. it has no chlorophyll and is entirely reliant on its host). It forms dense mats of twisted stems and penetrates the phloem vessels of its host plant. Pennings and Callaway (1996) studied how it affects interactions between salt marsh plants (pickleweed [*Salicornia virginica*], California sea lavender [*Limonium californicum*] and alkali heath [*Frankenia salina*]) in Carpinteria Salt Marsh in California). They recorded the biomass of communities in which dodder was present and absent, and how these interactions differed higher up (20 cm above mean high water mark) and lower down (17 cm above mean high water mark) the salt marsh. 'Large' patches of dodder were defined as being over 1 m^2 and comprised a central area of dead dodder which had killed its host plant, surrounded by a living circle of dodder. Patches of thick, live dodder covering less than 1 m^2 were described as 'small.'

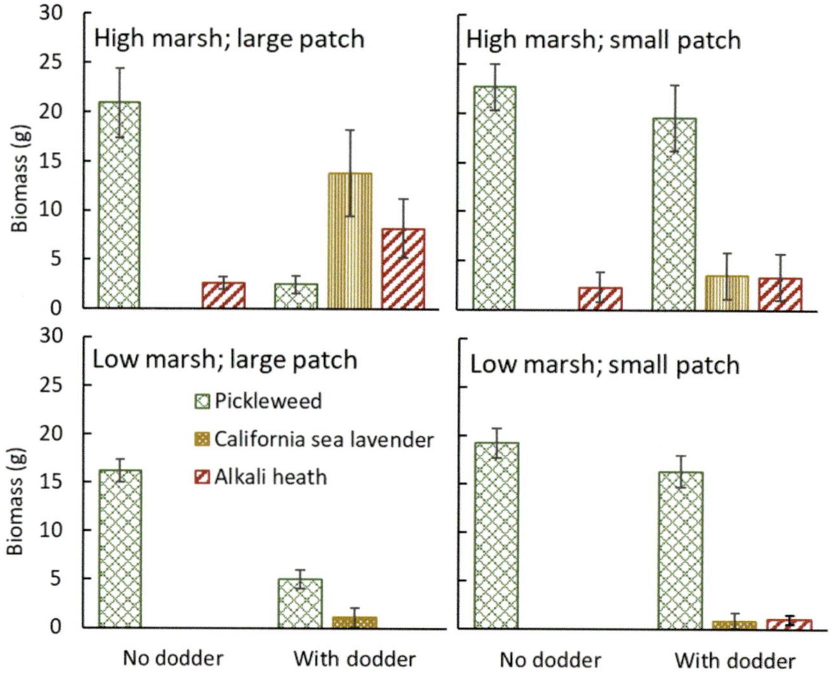

Figure 6.7 Mean biomass $\pm$ 1 SE of three salt marsh plants when parasitised by dodder or not, on the high and low marsh, and when dodder was in large or small patches. (Adapted from Pennings and Callaway, 1996. With permission of John Wiley & Sons – Books.)

Question 6.7.1: Use Figure 6.7 to work out which species is competitively dominant on the high and low marsh.

A. Dodder
B. Alkali heath
C. California sea lavender
D. Pickleweed

Question 6.7.2: Why don't small patches of dodder lead to increases in California sea lavender and alkali heath? (No more than 40 words are needed.)

Question 6.7.3: With which species would you predict that California sea lavender had the strongest positive correlation?

A. Dodder
B. Alkali heath
C. Pickleweed

Question 6.7.4: Why might alkali heath and California sea lavender have lower biomass on the low marsh even where there are large patches of dodder?

Question 6.7.5: Use the information and data given in this question to suggest how the presence of dodder could drive cycles of changes in vegetation and patchiness of species across the salt marsh. (Aim for around 100 words for a complete answer.)

Used with permission of John Wiley & Sons – Books; Pennings, S. C., Callaway, R. M. 1996. Impact of a Parasitic Plant on the Structure and Dynamics of Salt Marsh Vegetation. *Ecology*, 77(5), 1410–1419. Permission conveyed through Copyright Clearance Center, Inc.

Exercise 6.8: Avian vampire flies and their impact on Galapagos finches

The parasitic avian vampire fly *Philornis downsi* is native to South America. It was accidentally introduced to the Galapagos Islands in the 1960s. Adult flies feed on fruit and nectar but lay their eggs in birds' nests. The fly larvae feed on the nestlings' blood and have a severe detrimental effect on the young birds of some species. There is concern about the long-term impacts of this parasite on endemic bird populations. The closely related green warbler finch *Certhidea olivacea* (an insectivore) and small tree finch *Camarhynchus parvulus* (an omnivore) are affected differently by the vampire fly and have been studied to assess changes in impact over time (Table 6.8 and Figure 6.8).

Table 6.8 Impact of vampire flies in the nests of small tree finch and green warbler finch at incubation and nestling phases

			Percentage prevalence of vampire flies in nests				Mean number of larvae.nest^{-1} ± 1 SE	
			Incubation phase		Nestling phase			
Year	N Tree finch	N Warbler finch	Tree finch	Warbler finch	Tree finch	Warbler finch	Tree finch	Warbler finch
2012	37	46	24.9	80.6	100	100	37.4 ± 4.19	26.2 ± 2.9
2014	73	96	21.4	14.9	96	98	30.7 ± 2.62	24.6 ± 1.63
2015	54	49	11.9	50.7	93	94	39.1 ± 3.72	22.6 ± 2.5
2016	29	36	9.0	38.8	100	94	35.9 ± 4.54	25.0 ± 3.08
2017	36	72	0.0	59.7	83	85	30.6 ± 4.03	17.5 ± 2.07

Source: Adapted from Cimadom and Tebbich, 2021. Reproduced under the Creative Commons CC BY licence, copyright 2020 Cimadom and Tebbich.

Question 6.8.1: Overall, which species has the higher prevalence of parasitic fly larvae in the incubation phase?

A. Small tree finch
B. Green warbler finch
C. No difference between the species

Question 6.8.2: Overall, which species has the higher prevalence of parasitic fly larvae in the nestling phase?

A. Small tree finch
B. Green warbler finch
C. No difference between the species

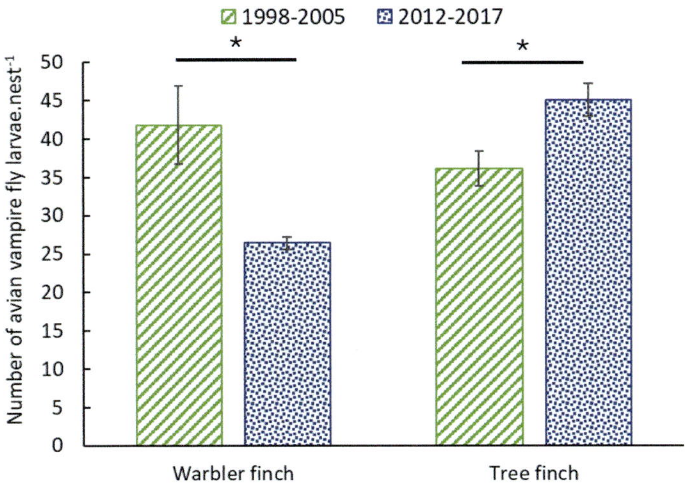

Figure 6.8 Mean number of avian vampire fly larvae in nests ± 1 SE in nests with nestlings over 5 days old. (Adapted from Cimadom and Tebbich, 2021. Reproduced under the Creative Commons CC BY licence, copyright 2020 Cimadom and Tebbich.)

Question 6.8.3: Overall, which species has the higher mean number of parasitic fly larvae in the nest?

A. Small tree finch
B. Green warbler finch
C. No difference between the species

Question 6.8.4: Which of these is likely to affect the host choice of avian vampire flies in these Galapagos finches? (More than one answer is correct.)

A. Host frequency
B. Host detectability
C. Tolerance of host—a 'good' parasite doesn't want to kill its host too fast
D. Likelihood of nest failure
E. Adaptation to host
F. Number of fly larvae already present in the bird's nest
G. 'Quality' of host blood

Question 6.8.5: It has been suggested that vampire flies find hosts using olfactory cues, in which case, the larger nests and nestlings of small tree finches (~13 g) should make it easier for the parasite to find these compared to green warbler finches (~8 g). From the data shown here, what makes this idea less likely to be the case? (No more than 40 words are needed.)

Question 6.8.6: The number of deaths of chicks is high and this occurs on average at 6 days for both species. Vampire fly larvae need a minimum of 6 days to develop to the pupal stage. Theory suggests that this should lead to the parasite selecting the host on which it can mature more quickly. What change in vampire fly behaviour might change due to this? (No more than 50 words are needed.)

7

Predation and herbivory

Exercise 7.1: Brown bear diet in Sweden

Brown bears (*Ursus arctos*) are omnivorous and in south-central Sweden are known to eat ants (*Formica* and *Camponotus* spp.), bilberries (*Vaccinium myrtillus*), crowberries (*Empetrum nigrum*), lingonberries (*Vaccinium vitis-idaea*) and newly born moose calves (*Alces alces*). Stable isotope analysis of hair samples taken from bears in spring can indicate the food eaten by bears over the previous summer. When bears have eaten more meat, there will be an increase in $\delta^{15}N$ in hairs, and when they have eaten more berries, there will be an increase in $\delta^{13}C$.

 DOI: 10.1201/9781003644095-7

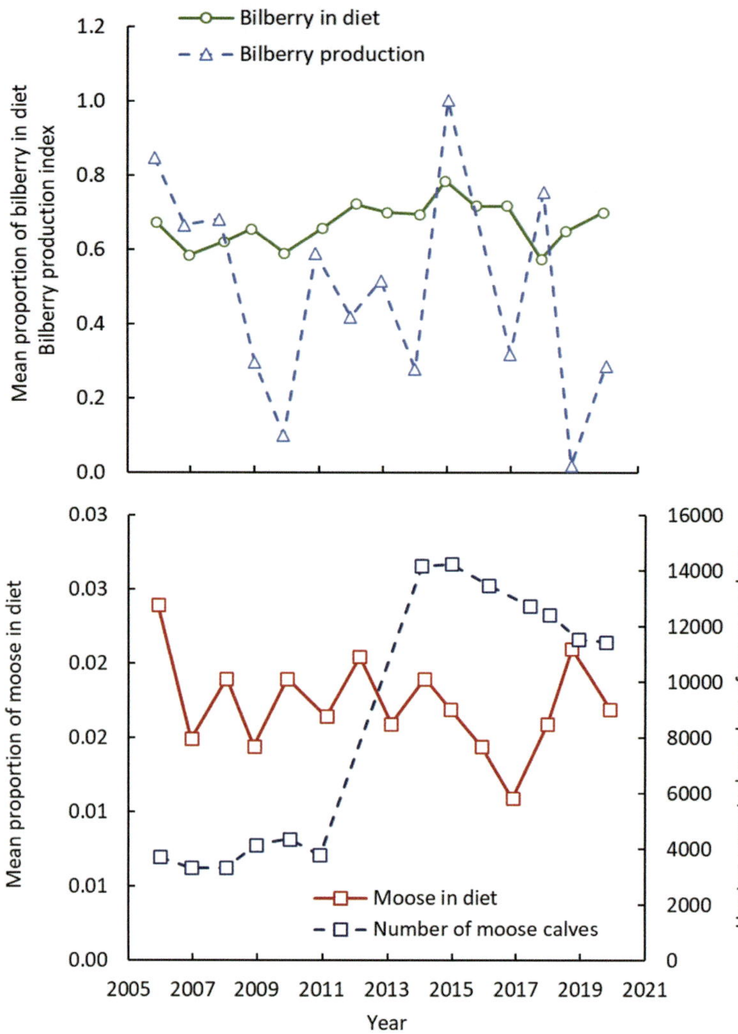

Figure 7.1 (Top) Proportion of bilberry in diet and bilberry production in the environment. (Bottom) The left-hand axis is the proportion of moose in bears' diets, and the right-hand axis is the number of moose calves reported by hunters. (Adapted from Mikkelsen et al., 2025. Reproduced under the Creative Commons CC BY licence, copyright 2025 Mikkelsen et al.)

The results shown in Figure 7.1 are from females with dependent offspring (solitary males and females show slightly different diets but the same broad patterns of feeding). The reason for the large increase in moose between 2011 and 2013 isn't clear but may relate to enhanced media coverage of moose numbers at that time, greater awareness amongst hunters and improved reporting of moose calves by hunters (A. J. Mikkelsen, pers. comm.).

Question 7.1.1: How did increased production of bilberries affect the amount of bilberries eaten by bears? (No more than 30 words are needed.)

Question 7.1.2: How did the increased availability of moose calves after 2011 change the proportion of moose in bears' diets? (No more than 20 words are needed.)

Question 7.1.3: Why might bears prefer to eat bilberries instead of newly born moose calves? (No more than 40 words are needed.)

Exercise 7.2: Population cycles of Arctic ground squirrels

The population dynamics demonstrated by snowshoe hares (*Lepus americanus*) and lynx (*Lynx lynx*) in the boreal forests of North America provide a classic textbook example of predator–prey cycles. These cycles can affect other predator and prey species in the same area. In the Kluane region of the southern Yukon, populations of Arctic ground squirrels (*Urocitellus parryii*) followed a similar cyclical pattern to snowshoe hares for over 30 years (Figure 7.2). Arctic ground squirrels live individually in burrows in open grasslands, tundra and boreal forests. In areas where there are no snowshoe hares, populations of ground squirrels do not show a cyclical pattern.

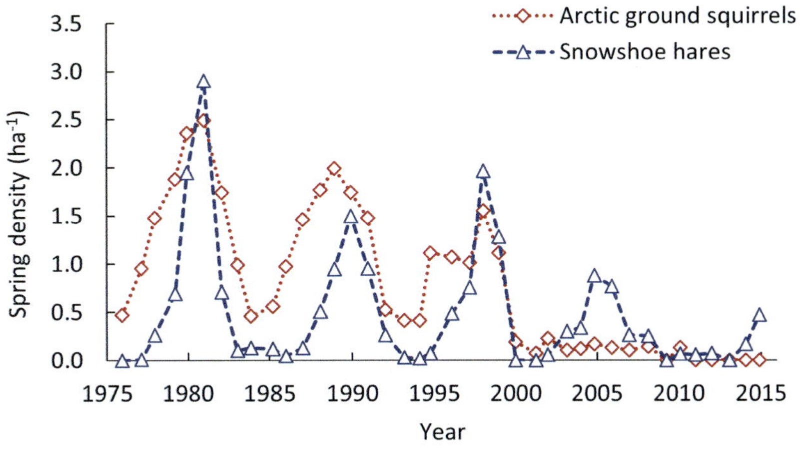

Figure 7.2 Spring density of Arctic ground squirrels and snowshoe hares from 1975 to 2015. (Adapted from Werner et al., 2016. Reproduced under the Creative Commons Attribution licence, copyright Werner et al. 2016.)

Question 7.2.1: What is the most likely reason for Arctic ground squirrel populations to decrease sharply approximately every 10 years?

A. Competition with snowshoe hares for food
B. Competition with snowshoe hares for territory

C. Apparent competition with snowshoe hares generated by a shared predator (lynx)
D. Predator satiation

Question 7.2.2: Give three reasons why the populations of ground squirrels failed to recover in the years after 2000? (No more than 80 words are needed.)

Exercise 7.3: Invasive mongooses in Japan

Small Indian mongooses (*Urva auropunctata*; previously in the genus *Herpestes*) were introduced to the Japanese island of Amami-Oshima in 1979 to control a poisonous pit viper. Mongooses have spread from the introduction point, and by 1999 there were estimated to be 5,000–10,000 individuals on the island.

Mongoose are generalist predators, and anecdotal evidence suggested that they were feeding on arthropods and some of the smaller endemic vertebrates on the island. Researchers sampled along a forest road to collect data on animal occurrence in relation to the length of time that mongooses have been present in an area (Figure 7.3).

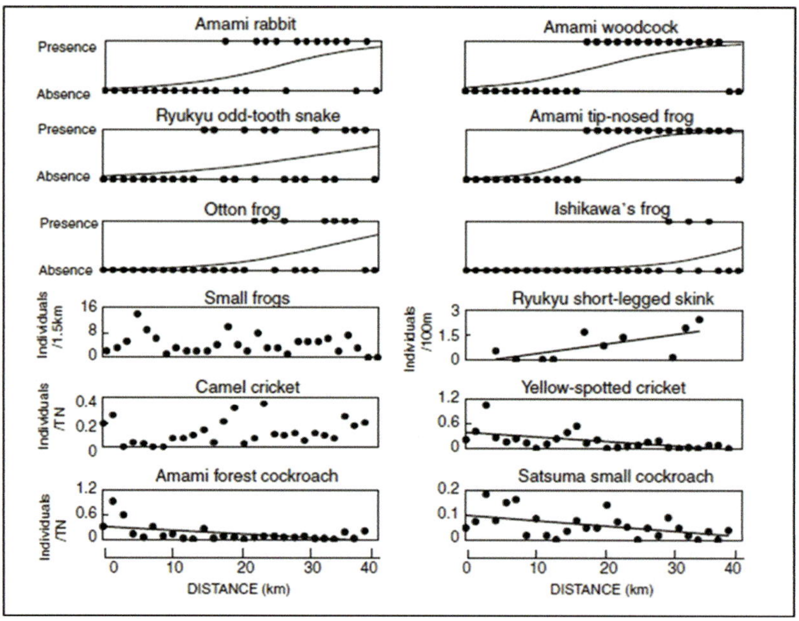

Figure 7.3 Occurrence (top graphs) and density (bottom graphs) of vertebrates and insects thought to be affected by mongooses. 'Distance' on the x-axis is measured from the release point of the mongooses. (From Watari et al., 2008. Reproduced with permission, copyright 2007 Springer Science Business Media, Inc.)

Question 7.3.1: Which species appear to have been negatively affected by the arrival of the mongoose on the island of Amami-Oshima?

A. Amami rabbit
B. Amami woodcock
C. Ryukyu odd-tooth snake
D. Amami tip-nosed frog
E. Otton frog
F. Ishikawa's frog
G. Small frog species
H. Ryukyu short-legged skink
I. Camel cricket
J. Yellow-spotted cricket
K. Amami forest cockroach
L. Satsuma small cockroach

Question 7.3.2: Do these findings fit with the anecdotal data that the mongooses were preying on arthropods and smaller vertebrates?

Question 7.3.3: What might explain the relationship between the presence of the mongoose and population size of yellow-spotted cricket, Amami forest cockroach and Satsuma small cockroach? (No more than 30 words are needed.)

Exercise 7.4: Skate and mussels in Strangford Lough

Strangford Lough is a sea loch in Northern Ireland where large elasmobranchs have recently been fished to extinction. Concurrently, populations of horse mussels (*Modiolus modiolus*) whose shells form reefs, also disappeared due to dredging for shellfish. Despite more than 10 years' protection from dredging, the mussel reefs have not recovered. Barrios-O'Neill et al. (2017) investigated the relationship between mussels, crabs (which eat mussels) and flapper skate (*Dipturus intermedia*; which eat crabs) in mesocosm experiments where young mussels were allowed to attach to a simulated reef (silicone-filled mussel shells) or to the bare base of the tank. Combinations of crabs and flapper skate were introduced to the tanks with and without the reefs; they measured predation on mussels and crab activity (Figure 7.4).

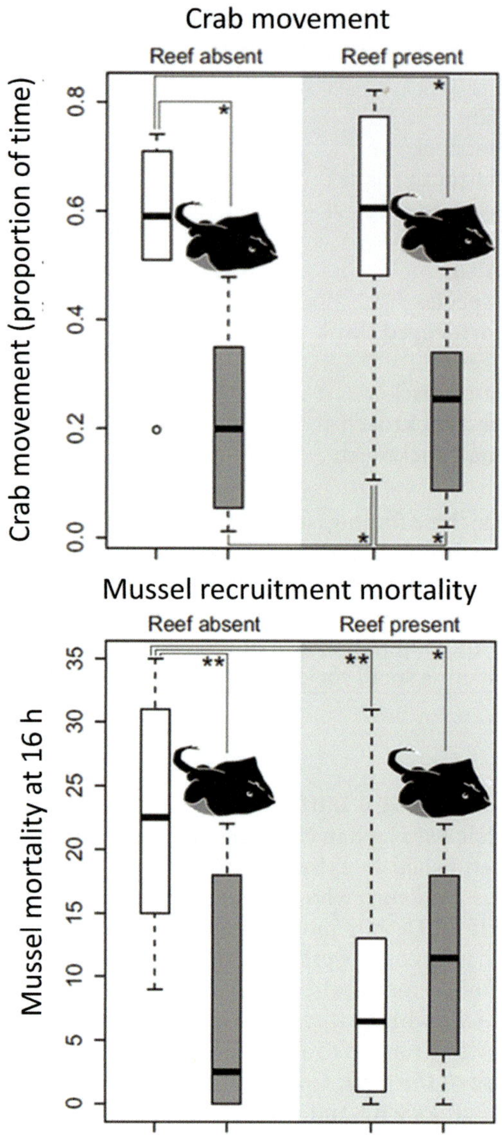

Figure 7.4 Crab activity (left) and mussel mortality (right) over 16 hours observation, in the presence (grey shading) and absence (white shading) of simulated reefs and in the presence (skate icon) and absence (no icon) of flapper skate. Treatments which were significantly different from each other are shown with asterisks: *$p < 0.05$; **$p < 0.01$. (From Barrios-O'Neill et al., 2017. Reproduced with permission, copyright 2017 Elsevier Ltd. All rights reserved.)

Question 7.4.1: Using Figure 7.4 and the preceding information, select the two statements from the following list that best explain the results found in this experiment.

A. Crabs are significantly more active on the simulated reefs than when the reef structure is absent.
B. Crabs are significantly less active on the simulated reefs than when the reef structure is absent.
C. Crabs are significantly less likely to be active when flapper skate are present regardless of the presence of reefs.
D. More mussels survive after 16 hours when flapper skate are present compared with when they are absent.
E. For mussels to be able to recolonise, either flapper skate or a reef structure need to be present.
F. For mussels to be able to recolonise, both flapper skate and a reef structure are necessary.

Question 7.4.2: Explain why this trophic cascade occurs. (No more than 50 words are needed.)

Question 7.4.3: How could mussel beds be reestablished in Strangford Lough? (No more than 20 words are needed.)

Exercise 7.5: Impact of invasive pythons

Burmese pythons (*Python bivittatus*; previously known as *Python molurus bivittatus*) have colonised large parts of Florida, following the release of unwanted pets and escape from a breeding facility during a storm. The pythons have been linked with severe declines in various common mammals such as raccoons, Virginia opossums, bobcats and rabbits.

Willson investigated whether changes in mammal populations affected freshwater turtles, a prey species of these mammals. He created artificial turtle nests using quail eggs and monitored predation using trail cameras (Figure 7.5a).

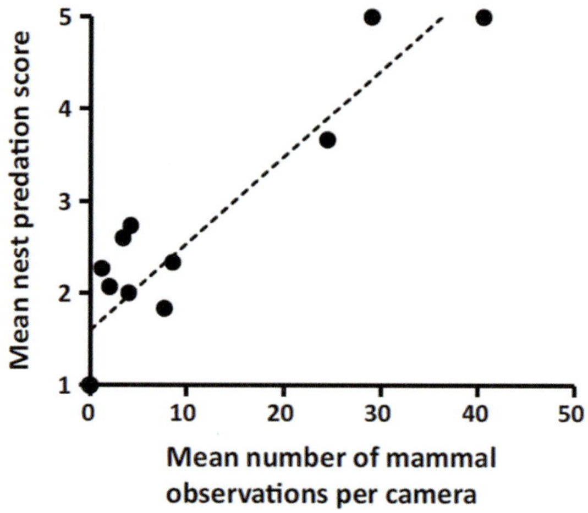

Figure 7.5a Relationship between predation scores for artificial nests and 'captures' of mammals on trail cameras. (From Willson, 2017. Reproduced with permission, copyright 2017 Willson. *Journal of Applied Ecology*, copyright 2017 British Ecological Society.)

Question 7.5.1: Why was it important for Willson to carry out this experiment before investigating the impact of pythons in more detail?

A. He needed to determine the relationship between the mammals and the pythons.
B. Because previous studies have shown that pythons and mammals compete for turtle eggs.
C. He needed to establish that mammals were the main predator of turtle nests.
D. Because pythons eat turtles.
E. Because pythons, turtles and mammals are all under pressure from habitat destruction in the Florida Everglades.

Question 7.5.2: Which of the following statements gives the most appropriate interpretation of these results?

A. There was a significant relationship between the number of mammals sighted and predation of artificial nests.
B. As the number of mammals observed increased, there were more predation events due to pythons eating the mammals.
C. There was a significant, positive relationship between the number of mammals sighted and the predation of artificial nests.
D. There was a positive relationship between the predation of artificial nests and the number of mammals sighted.

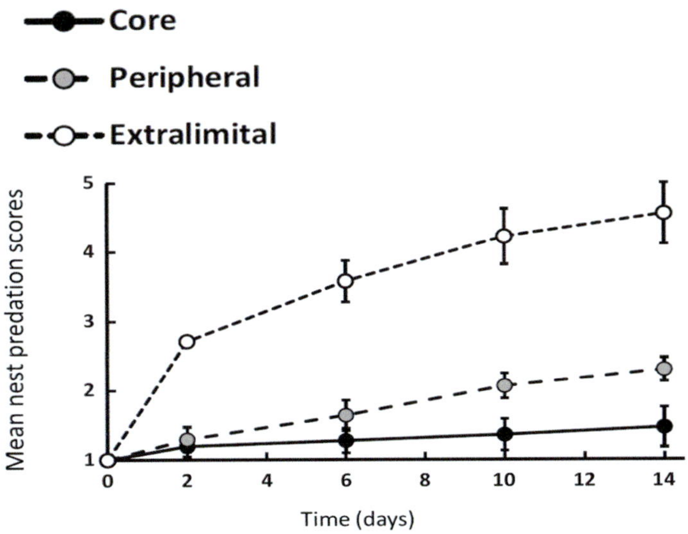

Figure 7.5b Mean nest predation scores means ± 1 SE for artificial 'turtle' nests containing quail eggs. 'Core' sites are where pythons have been established for the longest period of time. 'Peripheral' sites are areas where pythons have been documented more recently. 'Extra-limital' are where established python populations have not yet been confirmed. (From Willson, 2017. Reproduced with permission, copyright 2017 Willson. *Journal of Applied Ecology*, copyright 2017 British Ecological Society.)

Question 7.5.3: Using Figure 7.5b, which of these statements gives the most appropriate summary of Willson's findings?

A. Turtle nest predation is highest in the 'core' python areas because pythons are a significant predator of turtle eggs.
B. Turtle nest predation is highest in the 'extra-limital' python areas because pythons and mammals compete for resources.
C. Turtles are indirectly benefiting from the presence of pythons as there are fewer mammals, which predate turtle nests.
D. Turtles directly benefit from the presence of pythons because they protect them from predation by mammals.

Exercise 7.6: Predator–prey cycles

Lab cultures can be used to study interactions between prey and predator populations. Yoshida et al., (2003) used this approach to investigate cycling of algal cells (prey) and rotifers (predators). Figure 7.6 shows some of their results, demonstrating unusual predator–prey cycling. This effect only occurred when the researchers used a starting culture of algae that contained multiple genotypes. When they used algae derived from a single genotype, they observed a more typical predator–prey cycle. They suggest these findings are due to the evolution of algae that are resistant to rotifers.

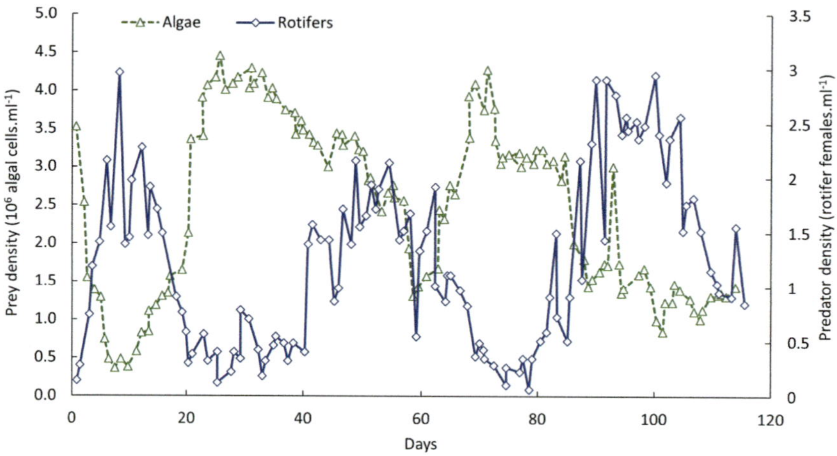

Figure 7.6 Predator (rotifers; right-hand axis) and prey (algae; left-hand axis) density over 118 days. (Adapted from Yoshida et al., 2003. Reproduced with permission, copyright 2003 Macmillan Magazines Ltd.)

Question 7.6.1: Which statement best describes the data in Figure 7.6?

A. When algae reach high densities, rotifer numbers increase showing a typical predator–prey cycle with a time lag between peaks.
B. High numbers of rotifers caused the algal populations to crash.
C. When algae populations are high, rotifers are low and vice versa.
D. Rotifers switch to different prey when algae are at low density.

Question 7.6.2: Why did this effect only appear when there were multiple genotypes of algae present? (Two answers are correct.)

A. Some genotypes are likely to be more resistant to grazing than others.
B. In a culture with mixed genotypes there will be less intraspecific competition for resources.
C. Rotifers select for more resistant algal genotypes.
D. Rotifers select for less resistant algal genotypes.

Question 7.6.3: In no more than 50 words, explain why populations of rotifers are able to increase as algal populations decline.

Exercise 7.7: Oak masting and weevil predation

Numerous tree species go through synchronised 'masting' events, where very large crops of seeds are produced in some years with very small crops in intervening times.

Question 7.7.1: What is thought to be the reason for this 'all or nothing' approach to seed production?

A. Predator satiation.
B. Tight coupling with climatic conditions.
C. After a very large production of seeds, trees need 'a few years off' to recover their resources.
D. It's a good survival strategy.

A research team in Spain looked at Mediterranean oaks (*Quercus* spp.) and acorn weevils (*Curculio* spp.; specialised seed predators) and the role of rainfall in determining masting events.

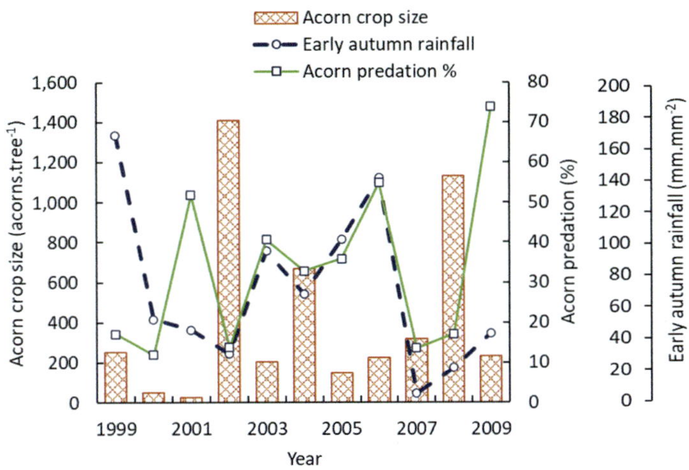

Figure 7.7a Acorn crop size (left-hand axis), bars; percentage acorn production (right-hand inner axis), green squares; early autumn rainfall (right-hand outer axis), blue circles. Data from a coppiced oak woodland in Collserola, eastern Spain. (Adapted from Espelta et al., 2017. Reproduced under the Creative Commons CC BY licence copyright 2017 Espelta et al.)

Question 7.7.2: Select the most appropriate explanation of the data shown in Figure 7.7a.

A. There is a very strong correlation between early autumn rainfall and acorn predation.
B. There is a very strong correlation between early autumn rainfall and masting.

C. Acorn predation tends to be greater in years of low acorn production.
D. There is no relationship between masting, acorn predation and early autumn rainfall.

At Huecas, another site in central Spain, the researchers recorded the weekly rainfall and the numbers of acorn weevils emerging (Figure 7.7b). There was a significant, positive correlation between the number of weevils emerging and the rainfall in the preceding 2 weeks ($r = 0.53$; $P < 0.001$).

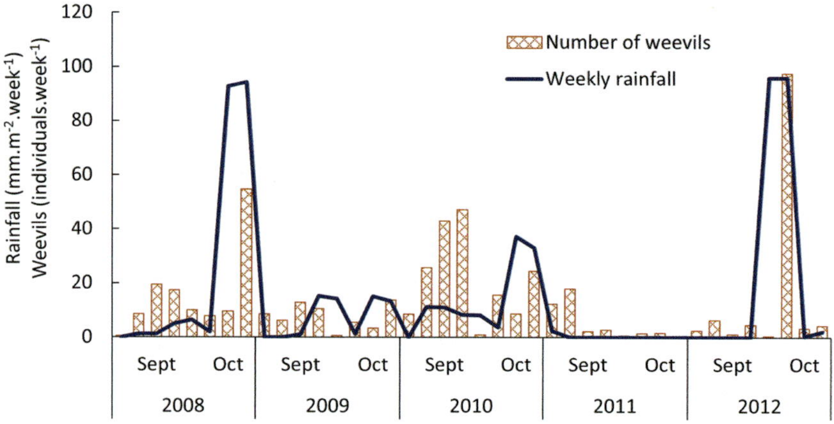

Figure 7.7b Weekly rainfall (blue line) and number of emerging weevils trapped weekly (bars) at Huecas between 2008 and 2012. (Adapted from Espelta et al., 2017. Reproduced under the Creative Commons CC BY licence copyright 2017 Espelta et al.)

Question 7.7.3: Figure 7.7a shows that the proportion of acorn predation in 2007 was similar to that in 2002 and 2008, both of which were mast years. Use Figure 7.7b to help explain this unexpected finding. (No more than 50 words are needed.)

Exercise 7.8: Impacts of predation on a tadpole community

Figure 7.8 shows the results from an experiment in North Carolina using artificial ponds where the numbers of the predatory eastern newt (*Notophthalmus viridescens*) were manipulated to investigate the impact of predation on the tadpoles of six species of frog and toad. Note that a number of these species have changed scientific names since this paper was published in 1983. Table 7.8 gives details.

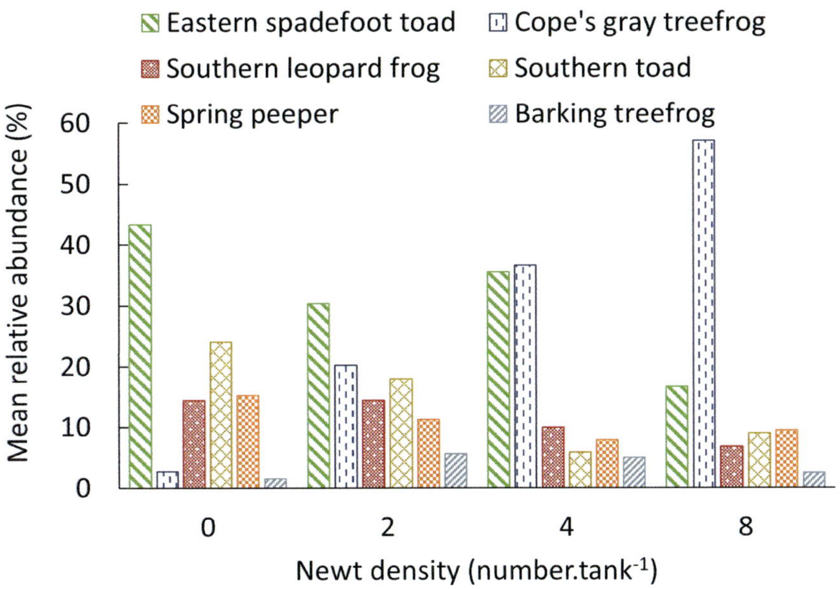

Figure 7.8 Numbers of successfully metamorphosed frog- and toadlets in relation to increasing numbers of predatory newts. Ponds contained the same initial densities of tadpoles. (Adapted from Morin, 1983. Reproduced with permission, copyright 1983 by the Ecological Society of America.)

Table 7.8 Old and new scientific names plus English names

Old scientific name	New scientific name	English name
Hyla chrysoscelis	*Dryophytes chrysoscelis*	Spring peeper
Hyla crucifer	*Pseudacris crucifer*	Cope's gray treefrog
Bufo terrestris	*Anaxyrus terrestris*	Southern toad
Hyla gratiosa	*Dryophytes gratiosus*	Barking treefrog
Rana sphenocephala	*Rana sphenocephala*	Southern leopard frog
Schaphiosus holbrooki	*Schaphiosus holbrooki*	Eastern spadefoot toad

Question 7.8.1: When no newts were present, which was the dominant species in this community?

Question 7.8.2: When high numbers of newts were present, which species was dominant?

Question 7.8.3: In no more than 60 words, state the interaction between Eastern spadefoot toad and Cope's gray treefrog. What adaptations would help Cope's gray treefrog to survive in the presence of Eastern newts?

8

Food webs and complex interactions

Exercise 8.1: Goldenrod, grasshoppers and spiders

Figure 8.1 shows changes in old field communities in northeast Connecticut when both the red-legged grasshopper (*Melanoplus femurrubrum*; an herbivore) and predatory nursery web spider (*Pisaurina mira*) were excluded, and when just the spiders were excluded. These grasslands are dominated by three 'plant functional groups': grasses, particularly smooth meadow grass (*Poa pratensis*); goldenrod (*Solidago rugosa*), a yellow-flowered, bulky dicotyledonous plant; and 'other herbs,' comprising a mixture of clovers, daisies, coneflowers, etc. (Table 8.1).

DOI: 10.1201/9781003644095-8

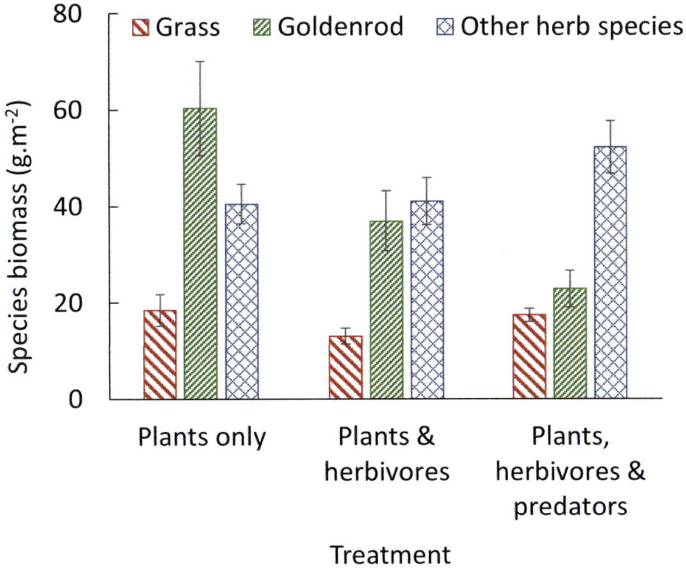

Figure 8.1 Biomass of each plant species when there is no herbivory, only herbivores, or herbivores and predators present. Bars show mean ± 1 SD. (Adapted from Schmitz, 2006. Reproduced with permission, copyright 1983 by the Ecological Society of America.)

Table 8.1 Results from Tukey post-hoc test for table showing comparisons between the biomass of functional group in each treatment

Response variable	Tukey's test results
Total community biomass	1 > (2, 3)
Grass biomass	(1, 3) > 2
Goldenrod biomass	1 > 2 > 3
Other herb species biomass	(1, 2) < 3
Functional group primary productivity[a]	1 > 2 > 3
Soil temperature	1 = 2 = 3
Soil moisture	1 = 2 = 3
Soil N mineralisation rate	1 < 2 < 3
Photosynthetically active radiation	1 < 2 < 3

Source: Adapted from Schmitz, 2006. Reproduced with permission, copyright 1983 by the Ecological Society of America.

Note: 1, plants only; 2, plants and herbivores; 3, plants, herbivores and predators. Numbers within brackets are not significantly different from each other.

[a]Data from Schmitz, 2003.

Question 8.1.1: Use the relevant sections of Table 8.1 to add in appropriate letters above the bars in Figure 8.1 to show which treatments are significantly different from each other within each functional plant group. Differences between each plant functional group and treatment were not reported; for clarity use the letters a, b and c (in whatever combinations are appropriate) for grasses; d, e and f for goldenrod; and g, h and i for other herbs. Note that you will not need to use all letters for all comparisons within each functional group.

Question 8.1.2: Which is the competitively superior plant functional group in the absence of herbivores?

A. Grasses
B. Goldenrod
C. Other herb species

Question 8.1.3: Which plant functional groups do grasshoppers prefer to eat? (More than one answer is correct.)

A. Grasses
B. Goldenrod
C. Other herb species

Question 8.1.4: What do grasshoppers do when spiders are present in the system?

A. Stop eating
B. Switch to grazing grasses
C. Switch to grazing goldenrod
D. Switch to grazing other herb species

Question 8.1.5: Which plants are responsible for the majority of primary production in this system?

A. Grasses
B. Goldenrod
C. Other herb species

Question 8.1.6: Goldenrod often forms extensive patches within old field systems. Use this additional information to explain the interactions between spiders, grasshoppers and their favoured types of food. (No more than 45 words are needed.)

Exercise 8.2: Grazing effects of red and roe deer

Deer populations have increased in many countries, largely due to the loss or decrease in numbers of top carnivores and reduced hunting pressure. Deer can have serious impacts on forest regeneration so it is helpful to understand their feeding choices. In the Netherlands, Ramirez and colleagues (2023) investigated habitat use by roe deer (*Capreolus capreolus*) and red deer (*Cervus elaphus*). Roe deer are solitary browsers (they feed on coarser twigs, shoots and leaves). Red deer live in herds and are described as 'mixed feeders'—they eat similar food to roe deer but also graze on grasses.

The researchers used trail cameras to record which deer were present in ten areas and what they were eating. The distance from the nearest town and to the edge of the forest was recorded for each plot. They also included hunting data, patch quality (percentage broadleaves, the remainder being less palatable conifers) and vegetation structure. Deer abundance and browsing intensity were calculated from the trail camera data. The data were analysed using linear mixed models and the results generated are shown in Figure 8.2.

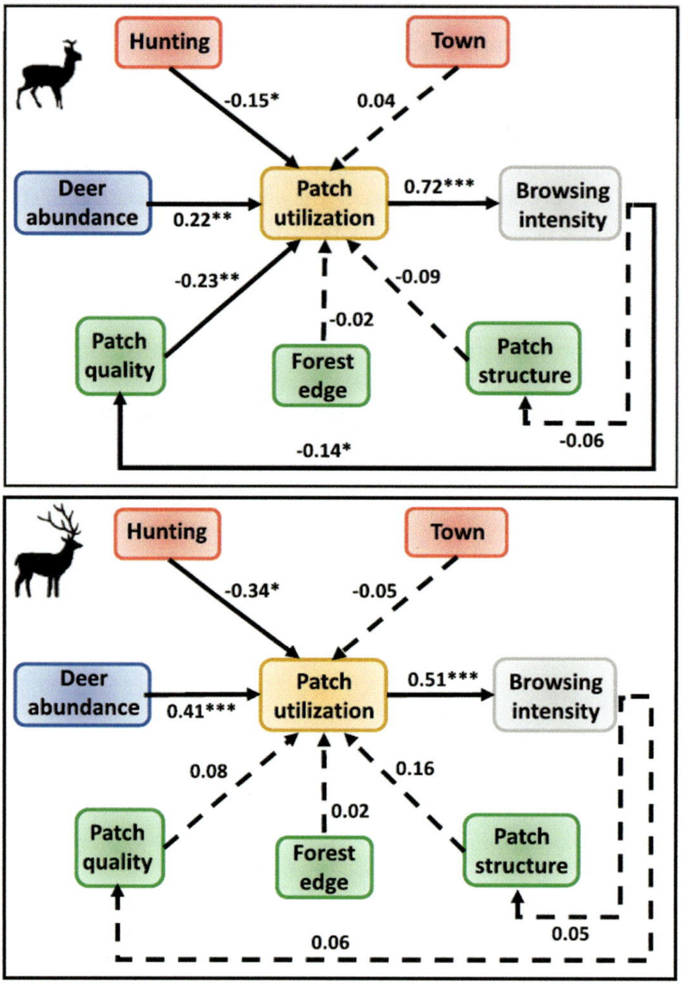

Figure 8.2 Path models based on results of linear mixed models for roe deer (top) and red deer (bottom). Numbers on the lines are the standardised coefficients of each relationship; these allow comparison of factors that have been measured using different scales. Larger numbers indicate that a relationship is more important; the sign shows whether the effect is positive or negative. The colour of boxes relates to the type of interaction. Dashed lines represent non-significant relationships; solid lines show significant relationships with asterisks next to the standardised coefficient: ***$P < 0.001$; **$P < 0.01$; *$P < 0.05$. (From Ramirez et al., 2023. Reproduced under the Creative Commons CC BY licence, copyright 2022 Ramirez et al.)

Question 8.2.1: Which of the measured factors are top-down forces?

A. Hunting
B. Town
C. Deer abundance
D. Browsing intensity
E. Patch quality
F. Forest edge
G. Patch structure

Question 8.2.2: Which of the measured factors are bottom-up forces?

A. Hunting
B. Town
C. Deer abundance
D. Browsing intensity
E. Patch quality
F. Forest edge
G. Patch structure

Question 8.2.3: Select the correct three statements from the following options.

A. Red deer experience top-down effects.
B. Red deer experience bottom-up effects.
C. Red deer do not experience either top-down or bottom-up effects.
D. Roe deer experience top-down effects.
E. Roe deer experience bottom-up effects.
F. Roe deer do not experience either top-down or bottom-up effects.

Question 8.2.4: Deer abundance was a stronger factor affecting patch utilisation for red deer than roe deer. Why might this be the case? (No more than 15 words are needed.)

Question 8.2.5: Why might patch quality be a significant factor for patch utilisation by roe deer but not by red deer?

A. Red deer live in herds and experience greater competition for food so patch quality is more important.
B. Roe deer have a more restricted range of suitable food as they only browse on twigs and stems.
C. Roe deer spend less time in high-quality patches as they can find the food they need more quickly.
D. Roe deer prefer woodland with a high proportion of conifers as there is increased cover from predators.

Question 8.2.6: What do the significant positive relationships between deer abundance and patch utilisation and between patch utilisation and browsing intensity show? (No more than 30 words are needed.)

Exercise 8.3: Lizards, homopterans, midges and shrubs

Interactions between species on Staniel Cay in the Bahamas have been studied to construct food webs. Lizards and spiders eat homopterans (grasshoppers and aphids) and gall midges. Spiders are eaten by lizards. Homopterans eat sea grape (*Coccoloba uvifera*), a type of shrubby tree that grows on beaches, leaving characteristic scars on leaves. Gall midges lay their eggs in the leaves and the larvae form distinctive galls. There are generally more scars than galls.

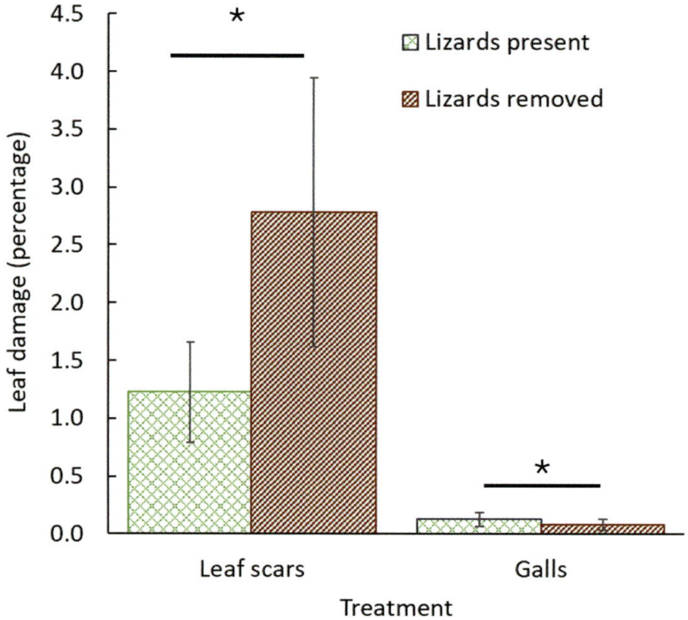

Figure 8.3a Percentage of leaf damage in the presence and absence of lizards. Bars show mean ± 1 SD. (Adapted from Spiller and Schoener, 1990. Reproduced with permission, copyright 1990 Springer Nature Limited.)

Question 8.3.1: Use the preceding information and the data in Figure 8.3a to choose the diagram in Figure 8.3b that gives the most appropriate overview of the interactions between these species. Thickness of arrows represents the per capita effect between the two species it connects (i.e. thicker arrows show a greater impact than thinner arrows).

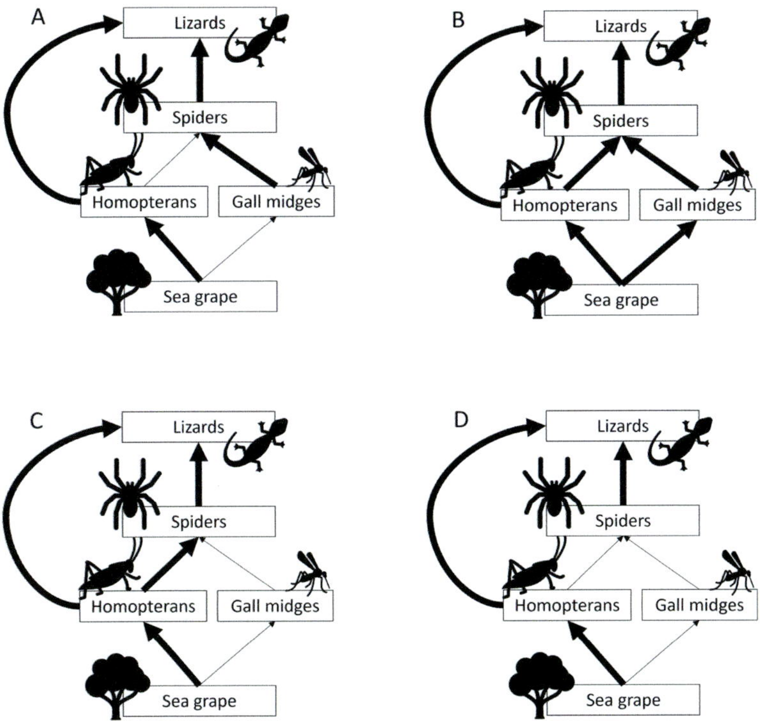

Figure 8.3b (Adapted from Spiller and Schoener, 1990. Reproduced with permission, copyright 1990 Springer Nature Limited.)

Exercise 8.4: Kruger food web

Here is a list of recent observations in Kruger National Park, South Africa. The statements are in no particular order.

- Wild dogs and cheetahs do less well in competition with lions and hyaena.
- Poachers believe carrying a lion body part when they hunt will protect them from capture by park rangers.
- Wild dogs are especially vulnerable to snares set for herbivores.
- Rhinos tend to keep open small patches (~4 m^2) of grass and this enhances the survival of invertebrates, small herbivores, ungulates and small birds.
- Homogeneous areas of vegetation are more vulnerable to hotter fires.
- Roan (antelopes) are a favoured prey of lions.
- Lions and spotted hyaena are common scavengers in Kruger.
- Rhino poachers hunt rhinos and take their horns.

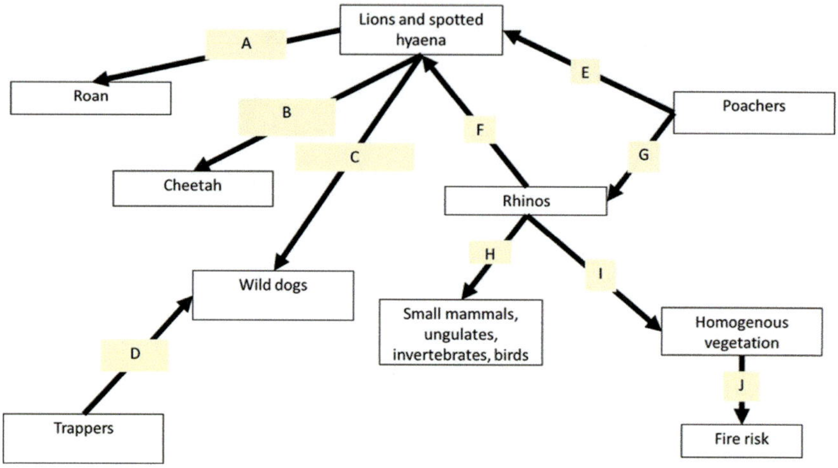

Figure 8.4 Interaction diagram from Kruger National Park. The impact is experienced by the species at the end of each arrow. (Adapted from Everatt et al., 2016. Reproduced with permission, copyright 2016 The Ecological Society of America.)

Question 8.4.1: Use the preceding statements to select the appropriate interactions from the following list for each letter in Figure 8.4.

- Increase
- Decrease
- Increase due to carcasses from poaching
- Decrease from competition
- Decrease due to snaring
- Decrease due to demand for body parts
- Decrease from predation

Exercise 8.5: Coastal food web

This exercise uses data from a 2-year study on the northwest coast of the US by Wootton (1992, 1993) involving various experiments to understand better the interactions between various organisms in this coastal community.

Goose barnacles (*Pollicipes* spp.) and blue mussels (*Mytilus trossulus*) colonise bare rock after storms have ripped off existing communities of algae and molluscs. Figure 8.5a shows the results of an experiment looking at the effect of caging areas of rock dominated by young goose barnacles. This excluded glaucous-winged gulls (*Larus glaucescens*) and American black oystercatchers (*Haematopus bachmani*), two of the key birds in this ecosystem.

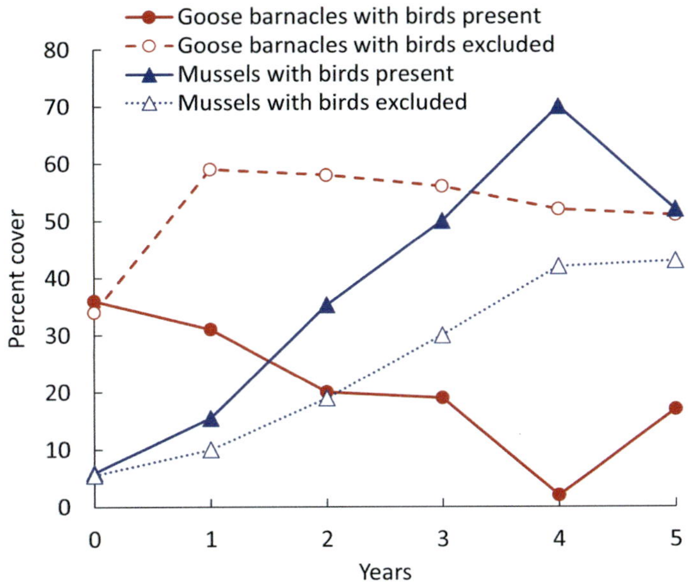

Figure 8.5a Percent cover of goose barnacles and mussels when birds were excluded or allowed access. Points show means. (Adapted from Wootton, 1993. Reproduced with permission, copyright 1993 by the Ecological Society of America.)

Question 8.5.1: Which of the following options gives the most appropriate explanation for this pattern of change in dominant species over time?

A. Mussels are the superior competitor and force goose barnacles off the rocks.
B. Goose barnacles are the superior competitor and prevent mussels from colonising bare patches.
C. Birds selectively prey on goose barnacles.
D. Birds selectively prey on mussels.
E. Goose barnacles are better colonisers than mussels.

Various species of limpet are also found in these coastal communities: ribbed limpet (*Lottia digitalis*), shield limpet (*L. pelta*) and checkered limpet (*L. strigatella*). Figure 8.5b shows what happened to the density of three of these species when birds were excluded.

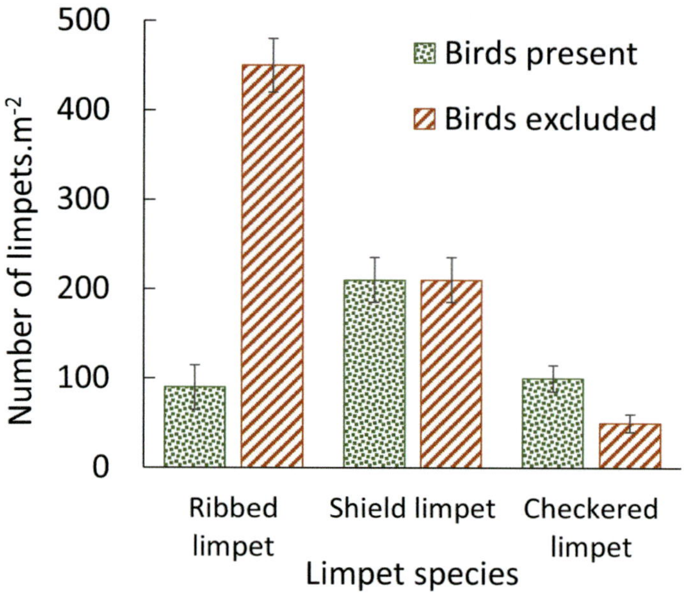

Figure 8.5b Change in cover of limpet species in the presence and absence of birds. Bars show mean ± 1 SE. (Adapted from Wootton, 1992. Reproduced with permission, copyright 1992 by the Ecological Society of America.)

Question 8.5.2: What is the most likely explanation for the increase in density of ribbed limpet in the absence of birds?

A. Birds eat ribbed limpets.
B. Birds eat shield limpets.
C. Birds eat checkered limpets.
D. Competition with shield limpets.
E. Competition with checkered limpets.

Question 8.5.3: What is the most likely explanation for the decrease in density of checkered limpet when birds are excluded?

A. Birds eat ribbed limpets.
B. Birds eat shield limpets.
C. Birds eat checkered limpets.
D. Competition with ribbed limpets.
E. Competition with shield limpets.

During his experimental work, Wootton observed that ribbed limpet and shield limpet were often attached to goose barnacles and mussels and their distribution did not appear to be random. To check this, he removed both limpet species from a set of mussels and goose barnacles, then reintroduced

40 of each limpet species into this area and protected it from predation with a cage. After 24 hours he counted where the limpet species had selected to settle. Figure 8.5c shows his results.

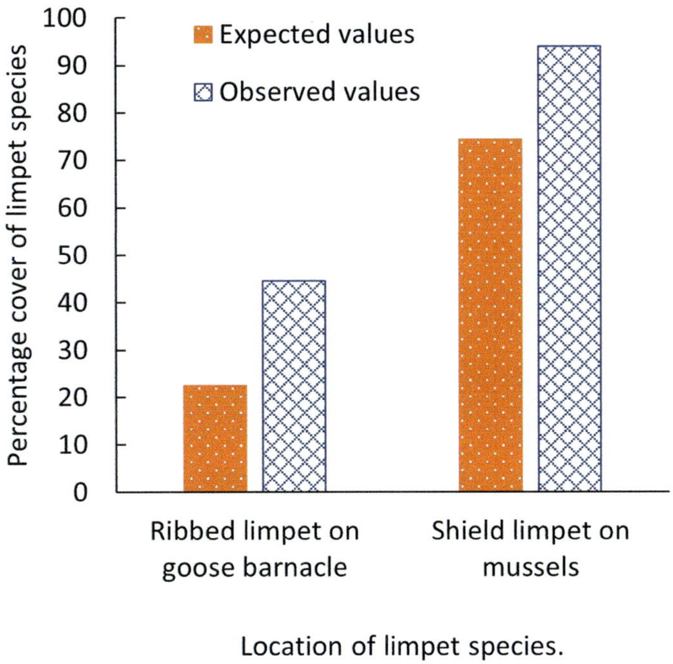

Figure 8.5c Settlement of ribbed and shield limpets on goose barnacles and mussels. Expected values were calculated from the area covered by barnacles and mussels, such that if barnacles covered 40% of the substrate, the expected proportion of ribbed limpet found there would be 40%. (Adapted from Wootton, 1992. Reproduced with permission, copyright 1992 by the Ecological Society of America.)

Question 8.5.4: Why might Wootton have observed this pattern? (No more than 15 words are needed.)

Question 8.5.5: Goose barnacles are pale coloured and mussels are dark coloured. What colour do you predict the limpet species to be? (Choose two options.)

A. Ribbed limpet is pale.
B. Ribbed limpet is dark.
C. Ribbed limpet is neutral.
D. Shield limpet is pale.
E. Shield limpet is dark.
F. Shield limpet is neutral.

Wootton also recorded the cover of algal species on the rock substrate and looked at the effect of the presence and absence of birds on these species. Figure 8.5d shows his findings.

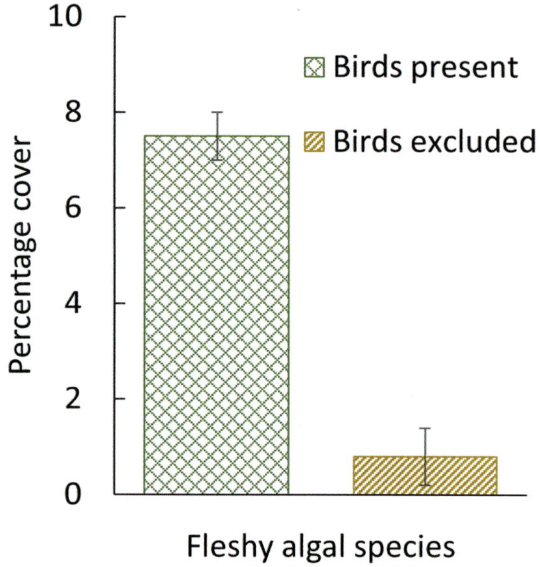

Figure 8.5d Percentage cover of fleshy algal species in the presence and absence of birds. Bars show means $\pm$ 1 SE. (Adapted from Wootton, 1992. Reproduced with permission, copyright 1992 by the Ecological Society of America.)

Question 8.5.6: Which of the following statements is most likely to explain what is happening in this ecosystem?

A. American black oystercatchers and glaucous-winged gulls eat algae; no birds lead to greater algal cover.
B. American black oystercatchers and glaucous-winged gulls eat algae; no birds lead to reduced algal cover.
C. Limpets eat algae; birds eat limpets; fewer limpets results in more algae.
D. Birds compete with limpet species.
E. Mussels and goose barnacles compete with algae for space; birds eat mussels and goose barnacles; fewer mussels and goose barnacles provide more substrate for algae to colonise.

Question 8.5.7: Which limpet is likely to have the biggest impact on fleshy algae?

A. Ribbed
B. Shield
C. Checkered

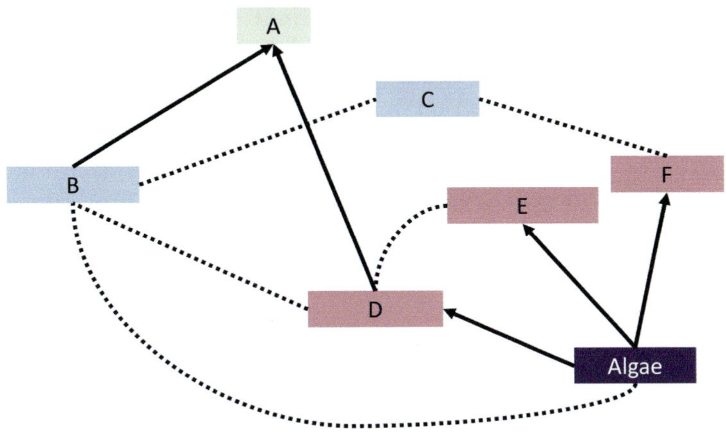

Figure 8.5e Connections between the main species in this coastal community. Solid arrows indicate direct interactions with the arrow showing direction of energy transfer, for example species D eats algae. Dotted lines show indirect interactions between species.

Question 8.5.8: Use the preceding information to select the appropriate species for each letter in Figure 8.5e. Select from birds, goose barnacles, ribbed limpet, shield limpet, checkered limpet, mussels.

Question 8.5.9: Now add the interactions into the diagram in Figure 8.5f. Choose from competitive exclusion, predation, exploits colour, herbivory, parasitism, mutualism (note: you may not need to use all of these options).

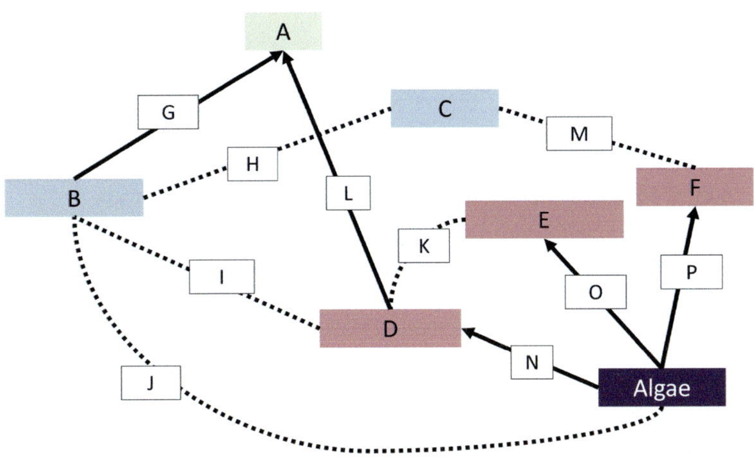

Figure 8.5f Connections between the main species in this coastal community and interactions. Solid arrows indicate direct interactions with the arrow showing direction of energy transfer, for example species D eats algae. Dotted lines show indirect interactions between species.

Exercise 8.6: Food webs and decomposition

Wenting and colleagues (2024) investigated the effects of different scavengers on trace elements in the soil below carcasses in the Netherlands. Using the carcasses of roe deer, they set up four treatments, as shown in Figure 8.6a: access by all birds and mammals; access by all birds and mammals except wild boar; access by all birds but no mammals; no access to birds or mammals. It was not practical to exclude invertebrates and microorganisms from any of the treatments.

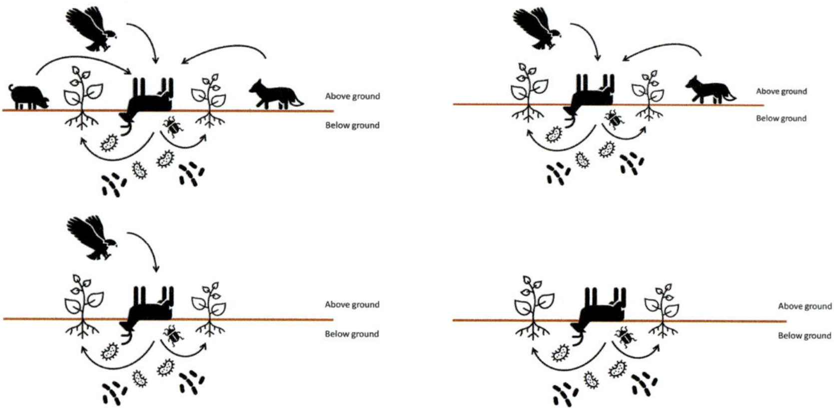

Figure 8.6a Schematic showing treatments for carcass decomposition experiment.

Question 8.6.1: Which of the following options is most likely to result from excluding all vertebrate scavengers?

A. Reduced movement of nutrients more widely across the landscape, creating high nutrient 'patches.'
B. Lack of vertebrate scavengers will reduce resources locally available to soil invertebrates as carcasses aren't opened up.
C. Areas around carcasses without vertebrate scavengers will become 'dead zones' where leachate from the dead animal kills vegetation.

The researchers measured changes in a range of elements over 40 weeks in plant shoots and roots adjacent to the carcasses and in the soil directly beneath the carcass. They analysed the concentrations of B, Cu, Mn, Mo, Ni, Zn, Ca, K, Mg, P and S. There were no significant differences between any of the treatments for any elements in plant roots or shoots or in the soil, except for Cu and Zn in soil (Figure 8.6b).

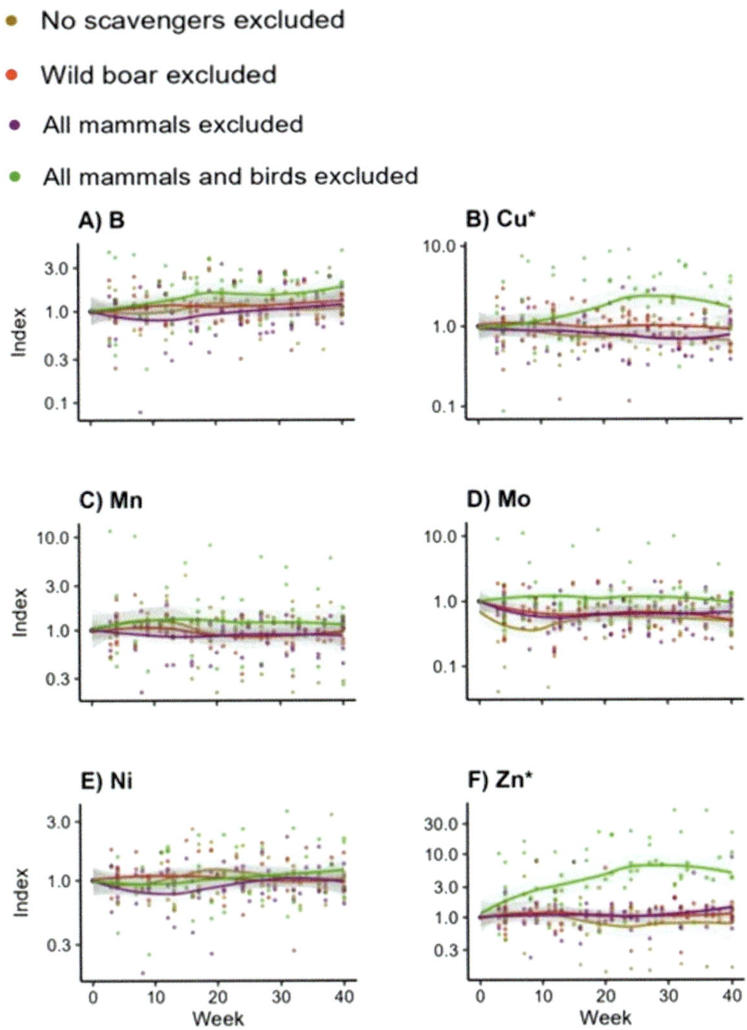

Figure 8.6b Changes in trace elements in the soil below the carcasses. The y-axis uses an index to show standardised concentrations with the starting point set as value 1 for all treatments. The asterisks indicate the elements which showed a significant result over the 40-week observations. (From Wenting et al., 2024. Reproduced under the Creative Commons Attribution 4.0 International License, copyright 2024 Wenting et al.)

Question 8.6.2: Looking at graph F in Figure 8.6b for Zn, what does this result actually mean? (Two answers are correct.)

A. In the absence of vertebrate scavengers, there was more Zn in the soil.
B. In the absence of vertebrate scavengers, there was less Zn in the soil.

C. In the treatment excluding mammals and the treatment excluding wild boar, there was no change in the concentration of Zn.
D. In the control treatment (no scavengers excluded), there was a reduction in Zn in the soil.

Question 8.6.3: Previous work has shown that wild boar speed up the rate of carcass decomposition. Why might this not have been the case here? (No more than 30 words are needed.)

Question 8.6.4: The researchers measured concentration of elements in plants growing around the carcasses but didn't find any differences between any of the treatments. What two additional measurements could they have taken to help explain this unexpected outcome? (No more than 50 words are needed.)

Question 8.6.5: What other group of invertebrates might be important to consider in this scavenger food web and why?

Exercise 8.7: Metabarcoding to unscramble water birds' diets

Fablet et al. (2024) collected bird faeces from eight wetland sites in northern France and used DNA metabarcoding to identify the plants and animals in their diet. They displayed their data using bipartite food webs (Figures 8.7a and 8.7b).

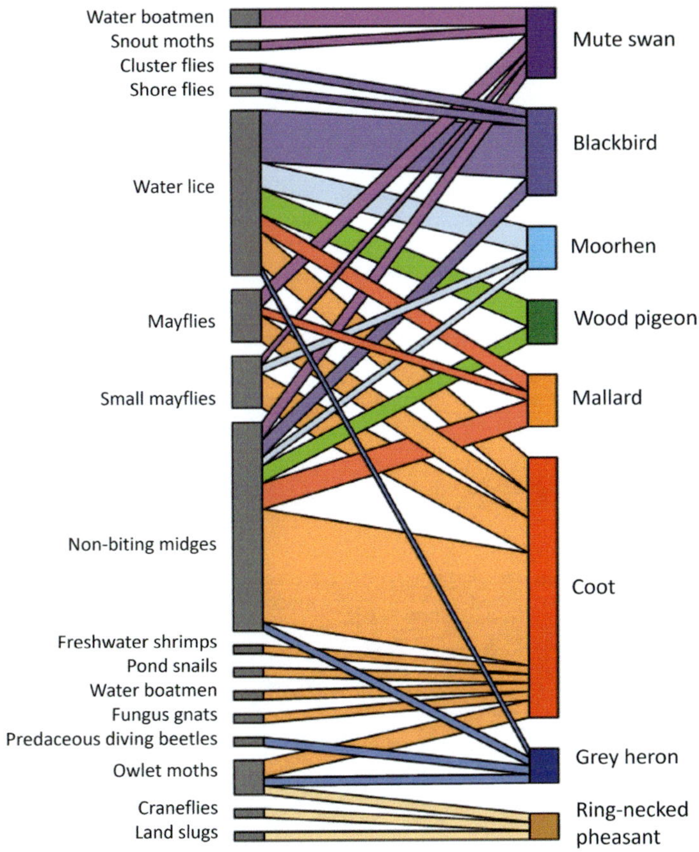

Figure 8.7a Bipartite food web of water bird diet of animal species. The right-hand coloured bars show the higher trophic level (birds) and the left-hand grey bars show invertebrate species. The length of the vertical bars is proportional to the total number of interactions involving these species. The diagonal/horizontal lines show connections between the trophic levels with the width of the line proportional to the number of interactions. (Adapted from Fablet et al., 2024. Reproduced under the Creative Commons CC BY ND licence, copyright © 2024 Fablet et al.)

Question 8.7.1: From which bird did the researchers collect the most samples?

A. Mute swan
B. Blackbird
C. Moorhen
D. Wood pigeon
E. Mallard
F. Coot
G. Grey heron
H. Ring-necked pheasant

Question 8.7.2: Which bird ate the greatest range of invertebrate families?

A. Mute swan
B. Blackbird
C. Moorhen
D. Wood pigeon
E. Mallard
F. Coot
G. Grey heron
H. Ring-necked pheasant

Question 8.7.3: Which invertebrate prey was most commonly eaten?

A. Water lice
B. Mayflies
C. Small mayflies
D. Non-biting midges
E. Owlet moths

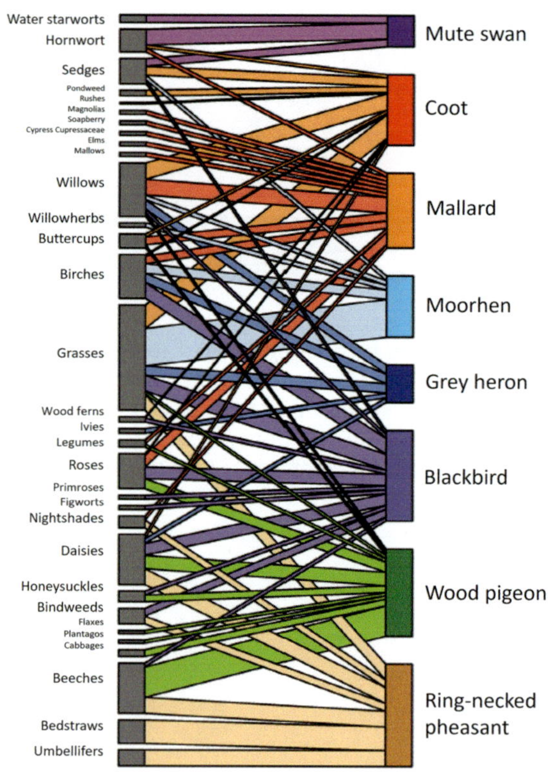

Figure 8.7b Bipartite food web of water bird diet of plant families. Other attributes as for Figure 8.7a. (Adapted from Fablet et al., 2024. Reproduced under the Creative Commons CC BY ND licence, copyright 2024 Fablet et al.)

Question 8.7.4: Herons mainly eat fish, molluscs, crustaceans and aquatic insects, but the metabarcoding results show they had consumed a range of plants, including willows, birches, grasses, ivies and daisies. What might explain this? (No more than 30 words are needed.)

Question 8.7.5: Which species of bird have the most generalist plant diet?

A. Mute swan
B. Blackbird
C. Moorhen
D. Wood pigeon
E. Mallard
F. Coot
G. Grey heron
H. Ring-necked pheasant

Question 8.7.6: Which three species of bird have the most specialised plant diets?

A. Mute swan
B. Blackbird
C. Moorhen
D. Wood pigeon
E. Mallard
F. Coot
G. Grey heron
H. Ring-necked pheasant

Question 8.7.7: Which food web (animals or plants) has the higher linkage?

Exercise 8.8: Response of soil food webs to defoliation

When outbreaks of autumn moth (*Epirrita autumnata*) and winter moth (*Operophtera brumata*) occur over short time periods, they can cause widespread defoliation and death of deciduous forests in northeastern Norway. Once trees have died, the vegetation becomes dominated by grass species, particularly wavy hairgrass (*Avenella flexuosa*), replacing the previous shrub layer of crowberry (*Empetrum nigrum*) and various other moorland dwarf shrub species (*Vaccinium* spp.). Crowberry is considered to be an allelopathic species as it produces leaf litter containing phenolic compounds.

A team of researchers investigated the impacts of the loss of trees on the soil food web by comparing transects along a gradient from strongly affected areas to undamaged woodland. They used environmental DNA metabarcoding and grouped species into trophic groups to create food webs (Figure 8.8a).

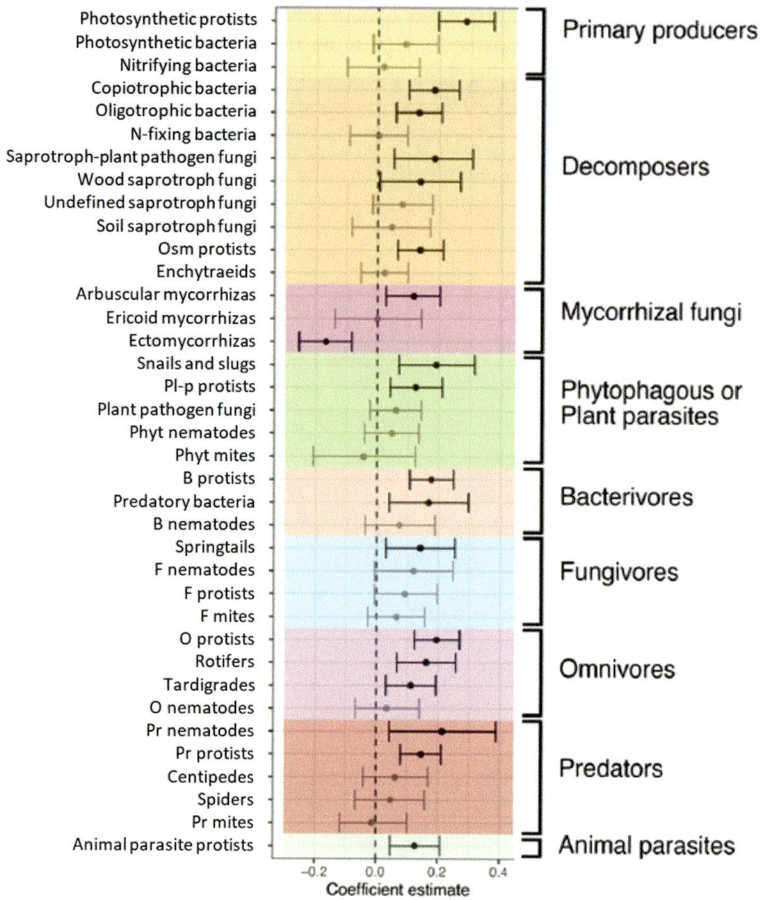

Figure 8.8a Coefficient estimates showing change in diversity of each decomposer group (i.e. Shannon diversity index) between forests with and without outbreaks of moths. A coefficient of zero shows no change, a positive coefficient shows an increase in moth-outbreak forests and a negative coefficient shows a decrease in moth-outbreak forests. Bars show 90% credible intervals (i.e. the probability that the true value falls within the interval). (Adapted from Calderón-Sanou et al., 2021. Reprinted under the Creative Commons CC BY licence, copyright 2021 Calderón-Sanou et al.)

Question 8.8.1: Some of the points and bars are shown in black and some in grey (see Figure 8.8a). What does the difference in colour show? (No more than 50 words are needed.)

Question 8.8.2: Overall, what effect did defoliation of these woodlands by moth larvae have on the diversity of soil organisms?

A. Increased diversity
B. Decreased diversity
C. Made no difference to diversity
D. None of the above

Question 8.8.3: Why might the diversity of ectomycorrhizal species have reduced while that of arbuscular mycorrhizas increased? (No more than 30 words are needed.)

Question 8.8.4: Why might wood-saprotrophic fungi have increased in diversity? (No more than 30 words are needed.)

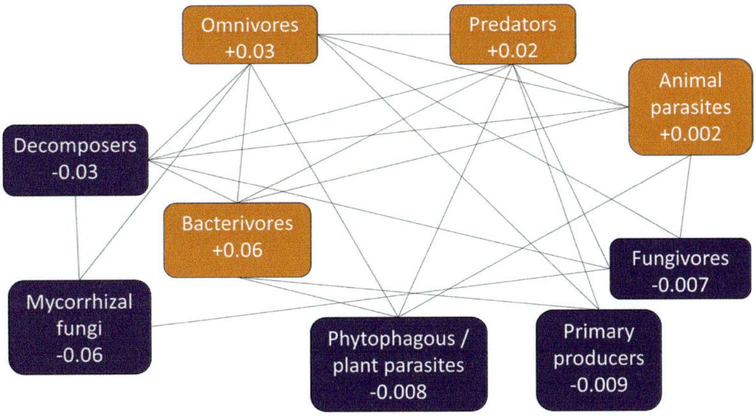

Figure 8.8b Links between trophic groups in the soil food web and how the relative abundance of each group changed from undamaged to defoliated woodlands. Orange boxes show an increase in relative abundance; those in purple show a decrease with numbers indicating the scale of change. (Adapted from Calderón-Sanou et al., 2021. Reprinted under the Creative Commons CC BY licence, copyright 2021 Calderón-Sanou et al.)

Question 8.8.5: Why might there be a reduction in the abundance of decomposers (–0.03) in forests affected by moth outbreaks? See Figure 8.8b. (No more than 30 words are needed.)

Question 8.8.6: The researchers hypothesised that the effect of defoliation and death of trees would decrease the trophic groups higher up the food web. Did their research support this hypothesis?

A. Yes
B. No
C. Partly

9

Impacts of climate change

Exercise 9.1: Birds on mountains

One response to a changing climate is for species to adjust their range to find conditions that are more suitable. Organisms can move northwards (in the Northern Hemisphere) or up mountains to remain in a similar climate to which they are adapted.

Question 9.1.1: What factors are likely to affect the ability of birds to shift altitude and successfully establish new populations? (Four suggestions are given in the answer, but there are potentially more than that!)

A study in Sweden and Norway used monitoring data from bird recording schemes between 1990 and 2020 and weather information from climate datasets, compiled for grid cells where there was a difference of 300 m altitude. The researchers compared the abundance of birds across the grids in study period 1 (1999–2002) with that in study period 2 (2015–2018). In study period 2, temperatures in March–April were higher and in May–June they were lower. There was no difference in July–August temperatures between the study periods. The results of the analysis are shown in Figure 9.1a.

DOI: 10.1201/9781003644095-9

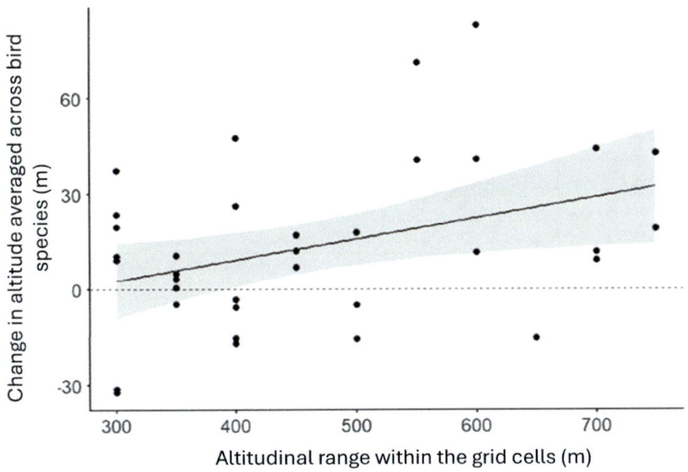

Figure 9.1a Change in altitude averaged across all bird species studied in relation to the altitudinal range in grid cells. Black points represent individual grid cells, the black line is the modelled relationship between abundance shift in altitude and altitudinal range within a grid cell, and the shaded grey area represents the 95% confidence interval around the modelled line. (From Couet et al., 2022. Reproduced under the Creative Commons CC BY licence, copyright 2022 Couet et al.)

Question 9.1.2: Which statement gives the most appropriate description of Figure 9.1a?

A. More birds moved to higher altitude areas (positive values on the graph) than lower altitude areas (negative values on the graph).
B. There is a greater abundance of birds at higher altitudes than at lower altitudes.
C. There are insufficient high-altitude areas to provide suitable habitat for birds that need to shift their range.
D. Where there is a greater altitudinal range in grid cells, birds showed a greater change in altitude.

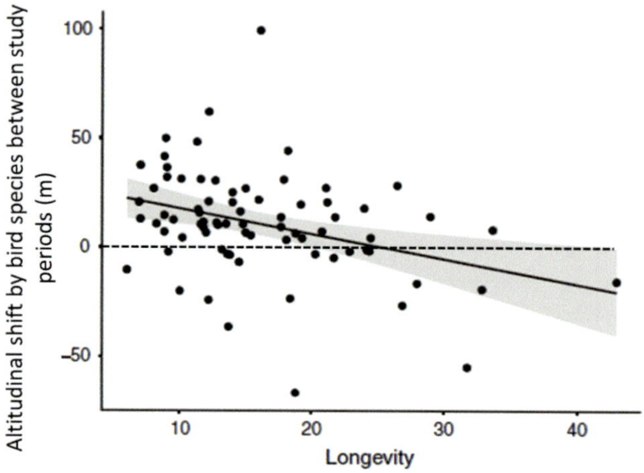

Figure 9.1b Longevity of bird species (years) plotted against the altitudinal distance moved by that species between the two study periods. Each point represents a bird species; the black line is the modelled line of fit and the grey shaded area shows the 95% confidence intervals around the modelled line. The dashed line shows no change in altitudinal range. (From Couet et al., 2022. Reproduced under the Creative Commons CC BY licence, copyright 2022 Couet et al.)

Question 9.1.3: What does the graph in Figure 9.1b show?

A. Short-lived birds showed a greater change in altitudinal range over time.
B. Long-lived birds showed a greater change in altitudinal range over time.
C. Most birds showed no change in altitudinal range over the two study periods.
D. Birds which already live at high altitude tend to be long-lived (K-selected), so there are reduced opportunities for them to increase their altitudinal range.

Question 9.1.4: What might explain the pattern seen in Figure 9.1b? (30 words are sufficient.)

Exercise 9.2: Parasitic trematodes' response to increased temperature

It is probable that relationships between parasites and their hosts will be affected by increasing temperature due to climate change, although the implications for individual species and communities are difficult to predict. Moving between hosts can be a vulnerable phase of development for many parasites, particularly if the developmental cycles use intermediate and definitive host species.

Trematode flatworms (flukes) are a very large group of parasites which generally use an aquatic snail as the intermediate host and a wide range of organisms as their final host. Large numbers of free-living cercariae are released from the intermediate snail host and seek their final host. Cercariae do not feed during this transmission phase and only live for up to 24 hours, so they need to find their next host quickly. Environmental conditions are likely to affect their ability to do this.

Selbach and Poulin investigated four trematode parasites that use the New Zealand mudsnail (*Potamopyrgus antipodarum*) as an intermediate host. *Maritrema poulini* and *Coitocaecum parvum* use crustaceans as their final host while *Apatemon* sp. and *Aporocotylid* sp. parasitise fish. The researchers ran cultures in the lab with mudsnails and the trematode species at 16°C (moderate summer water temperature) and 20°C (predicted under climate change scenarios) (Table 9.2).

Table 9.2 Production of cercariae, distance moved and survival time of trematodes infecting New Zealand mudsnails at different temperatures

	Mean change between 16°C and 20°C			
Species	Daily output snail^{-1}	Distance moved (%)	Survival time of cercariae (%)	Total active dispersal (m.24 hr^{-1})
M. poulini	+276	+41	–42	+205
C. parvum	+39	+3	–14	+23.4
Aporocotylid sp.	+73	–27	–7	+17.1
Apatemon sp.	+94	–1	–8	+76.7

Source: Selbach and Poulin, 2020. Reproduced with permission, copyright 2020 Springer-Verlag GmbH Germany, part of Springer Nature.

Question 9.2.1: What is the most likely reason that *M. poulini* had the greatest percent reduction in survival time?

A. The cercariae swam farther to try to find cooler water.
B. Increased temperature stimulated greater movement, but this is a non-feeding stage, so this activity probably reduced their energy reserves more quickly.
C. There was increased competition between cercariae for appropriate definitive hosts.
D. So many additional cercariae were produced that they were likely to be much smaller and weaker.

Question 9.2.2: Is the reduction in survival time likely to reduce the population size of the parasites? (No more than 20 words are needed.)

Question 9.2.3: Mudsnails and the small crustaceans that act as definitive hosts for *M. poulini* and *C. parvum* are highly abundant in the same habitats. Fish are much less common. What predictions would you make about the impacts of these four trematode species in warming environments? (Aim for 100 words for a full explanation.)

Question 9.2.4: What other data would help to predict the outcomes of increasing temperature on these parasites and their hosts? (Thirty words are sufficient.)

Exercise 9.3: Mycorrhizas and interactions

There is concern that the changing climate may disrupt mutualistic relationships such as the networks formed between mycorrhizal fungi and plants. Working in the field in northern Arizona, Gehring and collaborators (2014) investigated effects of herbivory, parasitism and competition on the mycorrhizal communities associated with pinyon pine (*Pinus edulis*) during drought and normal rainfall conditions. They used DNA sequencing to characterise the diversity of mycorrhizal communities when pinyon pine was affected by herbivorous scale insects or colonised by the parasitic plant mistletoe or experiencing competition against the shrub juniper for water and nutrients (i.e. below-ground competition). The data were analysed using non-metric multidimensional scaling (NMDS; see the Preface for details on how to interpret these types of output) (Figure 9.3a).

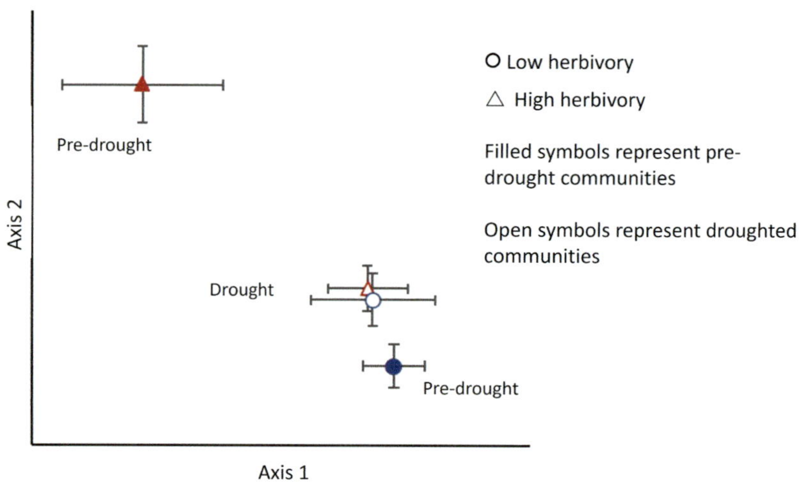

Figure 9.3a Non-metric multidimensional scaling (NMDS) of DNA sequences of mycorrhizal fungal communities categorised by water availability and level of herbivory. Points represent community centroids ± 1 SE. (Adapted from Gehring et al., 2014. Reproduced under the Creative Commons Attribution Licence, copyright 2014 Gehring et al.)

Question 9.3.1: What effect does drought have on the diversity of the mycorrhizal communities associated with pinyon pine?

A. Drought increases the impact of herbivory because aphids prefer dry conditions.
B. Drought reduces the impact of herbivory because aphids are less able to tolerate dry conditions.
C. The mycorrhizal communities have significantly less similarity.
D. The mycorrhizal communities have greater similarity.

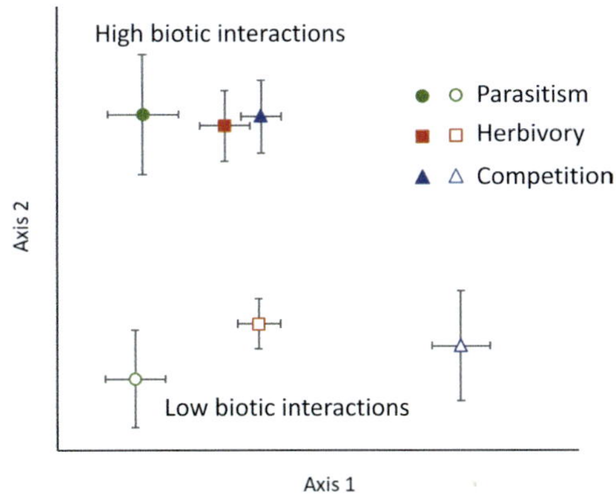

Figure 9.3b NMDS of DNA sequences of mycorrhizal fungal communities categorised by type of interaction (parasitism, herbivory or competition) and the intensity of the interaction (high or low). Open symbols represent low biotic interactions and filled symbols how high biotic interactions. Points represent community centroids ± 1 SE. (Adapted from Gehring et al., 2014. Reproduced under the Creative Commons Attribution Licence, copyright 2014 Gehring et al.)

Question 9.3.2: Using Figure 9.3b, what does Axis 1 appear to represent in this analysis? (No more than 70 words needed.)

Question 9.3.3: What does Axis 2 (Figure 9.3b) appear to represent in this analysis?

A. Intensity of biotic interaction
B. Degree of mycorrhizal colonisation
C. Response of pinyon pine to a range of stressors
D. Interaction between stressors

Question 9.3.4: Why might stressors such as drought, herbivory, parasitism and competition lead to the breakdown of the mutualism between plants and mycorrhizas? (No more than 40 words are needed.)

Question 9.3.5: What are the implications of these findings? (No more than 60 words are needed.)

Exercise 9.4: Damselfish interactions

Researchers in Australia looked at the behaviour of two species of damselfish (lemon damselfish [*Pomacentrus moluccensis*] and Ambon damselfish [*Pomacentrus amboinensis*]) at ambient CO_2 and elevated CO_2 and on unbleached and bleached coral. These species compete for space on reefs at the time that they settle out of the plankton (Figure 9.4a).

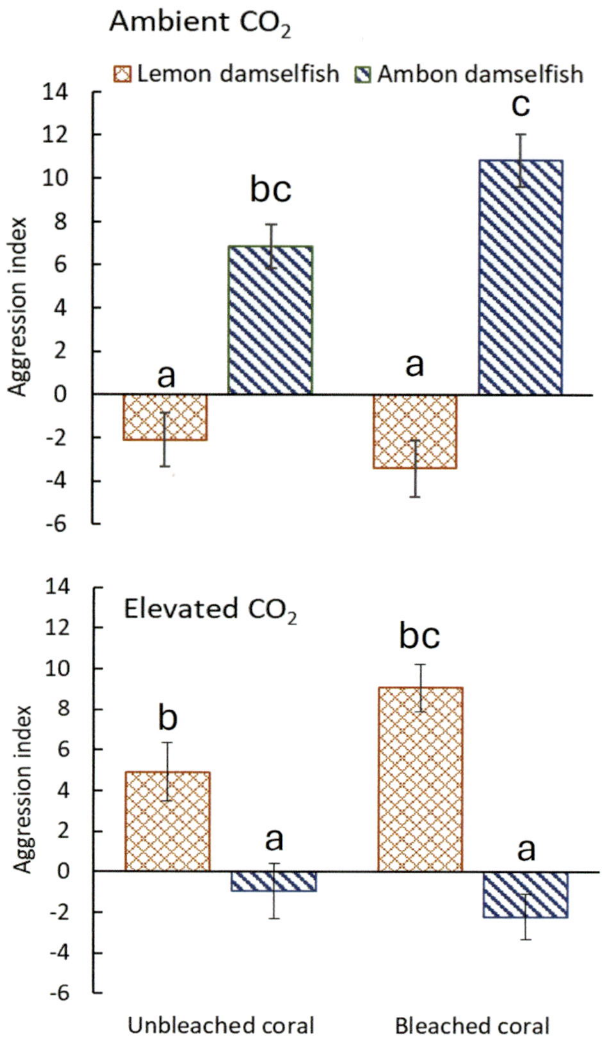

Figure 9.4a Aggression scores for lemon and Ambon damselfish at ambient and elevated CO_2 and on bleached and unbleached reefs. Bars show mean ± 1 SE. Bars sharing a letter are not significantly different from each other. (Adapted from McCormick et al., 2013. Reproduced with permission, copyright 2013 McCormick et al.)

Question 9.4.1: Under conditions of ambient CO_2, which species is the dominant competitor?

A. Lemon damselfish.
B. Ambon damselfish.
C. Neither; it varies between the two species.
D. They form a mutualism.

Question 9.4.2: Which is the dominant competitor under conditions of elevated CO_2?

A. Lemon damselfish.
B. Ambon damselfish.
C. Neither; it varies between the two species.
D. They form a mutualism.

Question 9.4.3: How do the interactions between the damselfish change between bleached and unbleached reefs?

A. There is a greater effect on bleached corals for both species.
B. There is a smaller effect on bleached corals for both species.
C. There is no difference in aggression on bleached and unbleached corals.
D. There is a significant statistical interaction showing that the species respond differently to coral bleaching.

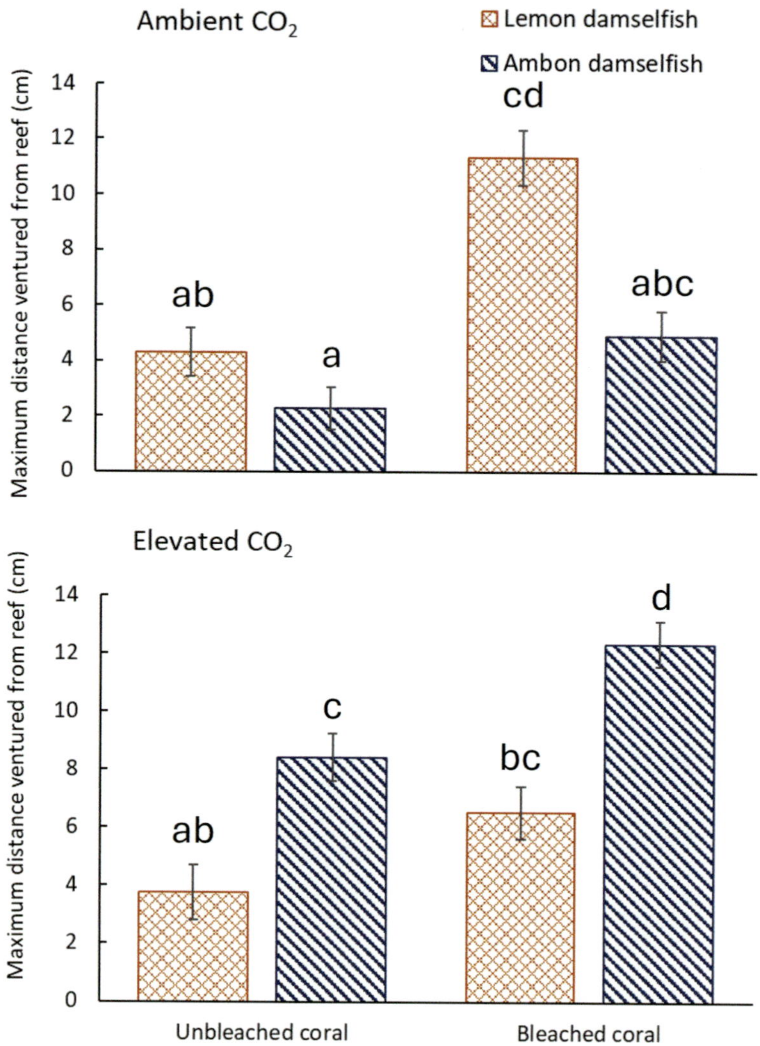

Figure 9.4b Distance moved from the reef at ambient and elevated CO_2 and on bleached and unbleached reefs. Bars show mean $\pm$ 1 SE. Bars sharing a letter are not significantly different from each other. (Adapted from McCormick et al., 2013. Reproduced with permission, copyright 2013 McCormick et al.)

Question 9.4.5: In no more than 60 words, explain how the behaviour of these damselfish changes in relation to the distance they move away from the reef when CO_2 concentrations increase. See Figure 9.4b.

Question 9.4.6: In no more than 75 words, explain how elevated CO_2 might impact this community of reef fish.

Exercise 9.5: Eider in the Arctic

Eider (*Somateria mollissima*) are a long-lived sea duck that breeds in northern Europe up to the High Arctic. They feed on molluscs, which they dive to collect, swallow whole and crush within their muscular gizzard. In Svalbard, populations have been monitored on two sites. On one site, where eider nests are collected for down, egg predators (Arctic fox, polar bear, Arctic skua and glaucous gull) were shot or chased away. On the other site within a nature reserve ('control'), there was no predator removal. Data from over 20 years were analysed to assess the effects of climate change and predation on populations of eider (Figure 9.5a).

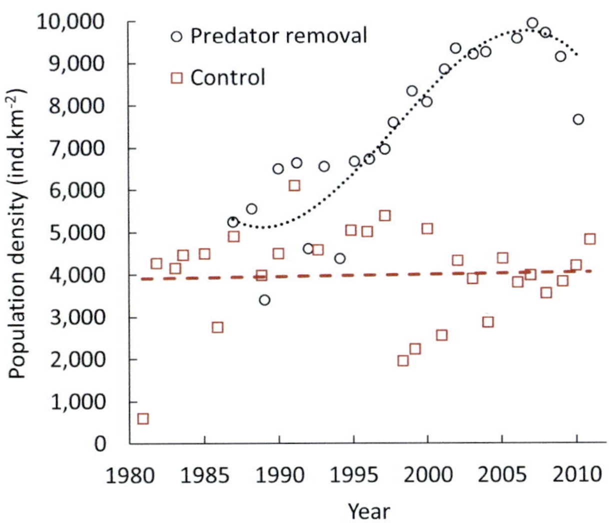

Figure 9.5a Population density of eider ducks from 1980 to 2012 on sites with and without predator removal. (Adapted from Hanssen et al., 2013. Reproduced under the Creative Commons CC BY licence, copyright 2013 Hanssen et al., *Ecology and Evolution* published by John Wiley & Sons Ltd.)

Question 9.5.1: What effect did predator removal have on the eider population?

A. Eider populations increased in the absence of predators.
B. Eider populations decreased in the absence of predators.
C. Predators made no difference to the population size of eiders, suggesting they are limited by other factors such as competition for food.
D. There is insufficient data to assess the impact of predators on eider populations.

The researchers analysed the population density using a linear model including parameters to assess the effects of predators, climate (North Atlantic Oscillation gives an indication of large-scale winter climate) and size of the eider population the previous year ('lagged' population size) (Table 9.5).

Table 9.5 Results of linear model to assess the impact of predators and climatic factors on eider populations

Parameter	Estimate	Standard error	*t*-Value	*P*
Predator removal	0.580	0.125	4.630	<0.001
Lagged July temperature (2 years previously)	0.066	0.048	1.360	0.181
North Atlantic Oscillation (winter)	−0.015	0.019	−0.773	0.444
Ice concentration (April)	−0.080	0.003	−2.573	0.014
April temperature	−0.012	0.011	−1.118	0.270
Lagged eider population size	−0.102	0.184	−0.554	0.583
Lagged eider population size × predator control	0.792	0.218	3.633	0.001

Source: Adapted from Hanssen et al., 2013. Reproduced under the Creative Commons CC BY licence, copyright 2013 Hanssen et al., *Ecology and Evolution* published by John Wiley & Sons Ltd.

Question 9.5.2: Did predator removal have a positive or negative effect on eider population size?

Question 9.5.3: The results from the analysis showed that ice concentration in April was a significant factor affecting the size of eider populations. Choose the two correct statements from the following list.

A. More ice in April increased eider populations.
B. More ice in April decreased eider populations.
C. Warming climate is likely to negatively impact eider populations.
D. Warming climate is likely to positively impact eider populations.

Question 9.5.4: How might the presence or absence of sea ice affect eider? (No more than 30 words are needed.)

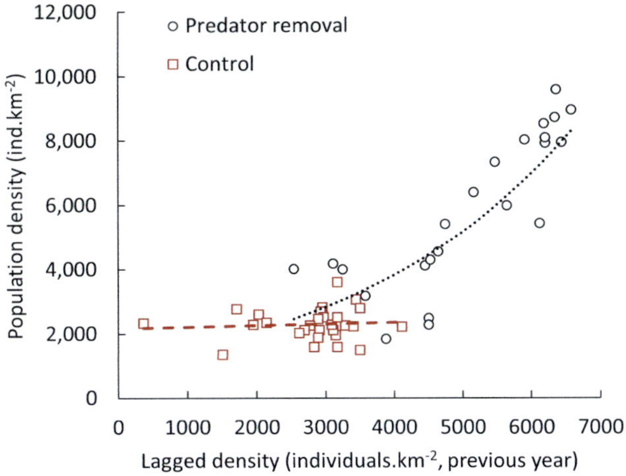

Figure 9.5b Population density in relation to lagged population density from the previous year for populations with and without predator removal. (Adapted from Hanssen et al., 2013. Reproduced under the Creative Commons CC BY licence, copyright 2013 Hanssen et al., *Ecology and Evolution* published by John Wiley & Sons Ltd.)

Question 9.5.5. Use Figure 9.5b to explain why the analysis (Table 9.5) showed a significant interaction between lagged eider population density and predator removal. (One hundred words will give a complete answer.)

Exercise 9.6: Warming and drying in tropical forest soils

Small amounts of ^{14}C exist naturally in the atmosphere. Nuclear weapons testing in the 1950s–1960s doubled this amount. Weapons testing was banned in 1963 and atmospheric ^{14}C declined from this time as it was taken up by plants and absorbed by the oceans. Measuring ^{14}C of CO_2 respired by soils allows estimation of the 'age' of the carbon substrates being used as an energy source (i.e. from decades to a few years ago). A team of researchers used this approach to investigate the effects of experimentally warming soils in lowland tropical rainforest in Panama and recorded the efflux of CO_2 from the soil and the amount of ^{14}C it contained (Figure 9.6).

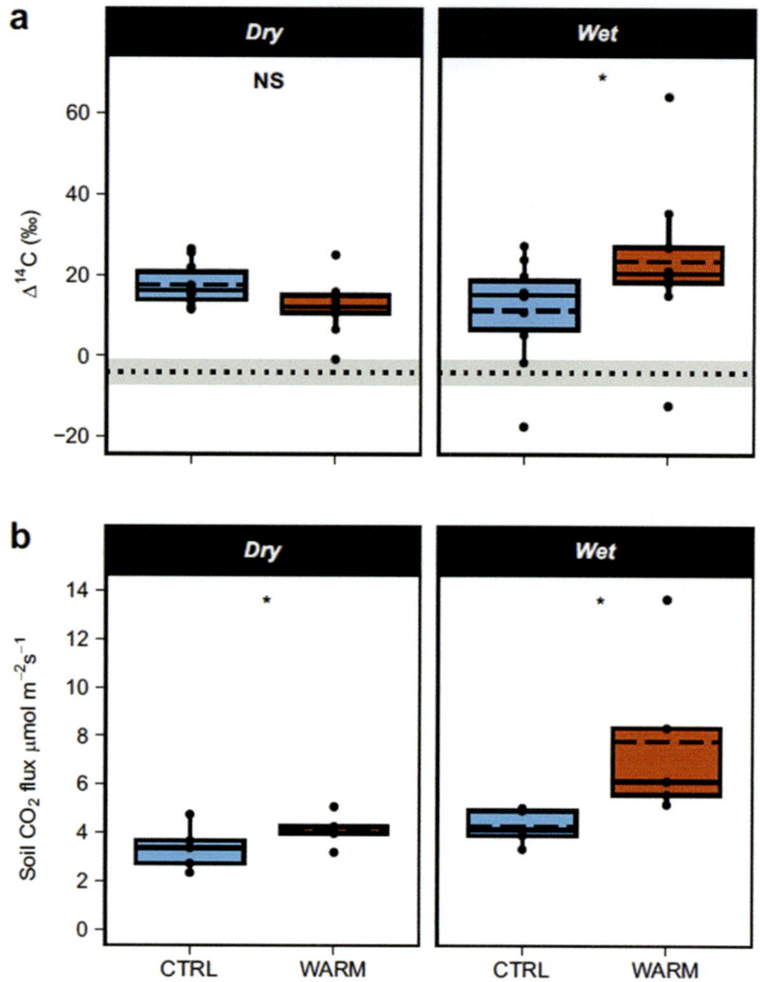

Figure 9.6 (a) Concentration of ^{14}C in wet and dry soils under ambient or warmed conditions. $\Delta^{14}C$ is the difference in ^{14}C between the measured concentration and the current concentration in the atmosphere. (b) Soil efflux of CO_2 in wet and dry soils under ambient or warmed conditions. Solid lines show medians, dashed lines show means; the boxes show the upper and lower quartiles (Q1 and Q3); whiskers show minimum and maximum ranges, and solid points show observations. Asterisks indicate a statistically significant difference between the control and the warming treatment. The dotted line shows the average atmospheric ^{14}C in this region ± 1 SE (grey shading in (a)). (From McFarlane et al., 2024, reproduced under the Creative Commons CC BY licence, copyright 2024 McFarlane et al.)

Question 9.6.1: What else has been happening since the 1960s that would reduce the ratio of ^{14}C:^{12}C in the atmosphere? (No more than 20 words are needed.)

Question 9.6.2: Figure 9.6 shows the effect of soil warming (WARM) in dry and wet seasons. Soils were warmed by burying resistance cables at 1.2 m depth to warm the entire soil profile by around 4°C. The warming treatment ran for 18–24 months before samples were taken for this experiment. Select the correct statements from the following list.

A. Warming increased the soil output of CO_2 in the dry season.
B. Warming increased the soil output of CO_2 in the wet season.
C. There was no difference in the output of CO_2 between the control and warmed plots in the dry season.
D. Warming decreased $\Delta^{14}C$ in the dry season.
E. Warming decreased $\Delta^{14}C$ in the wet season.
F. Warming increased $\Delta^{14}C$ in the dry season.
G. Warming increased $\Delta^{14}C$ in the wet season.

Question 9.6.3: Warmed soils respired greater amounts of ^{14}C in the wet season. Why is this relevant? (No more than 30 words are needed.)

Question 9.6.4: Roots usually respire CO_2 with ^{14}C values close to the current atmospheric level, so from where were the higher amounts of ^{14}C in warmed soils in the wet season actually coming? (No more than 20 words are needed.)

Question 9.6.5: What are the implications of these findings? (No more than 30 words are needed.)

Exercise 9.7: Arctic food webs

Capelin and Arctic cod are important 'forage fish'; they are responsible for the transfer of nutrients from phytoplankton and zooplankton to predators. Capelin have been moving northwards in response to warming seas, while Arctic cod tend to decrease in abundance where waters are warming. In Cumberland Sound, Nunavut, Canada, researchers used stable isotope analysis to investigate changes in diet from samples collected between 1990–2002 and 2005–2012 (Figure 9.7a).

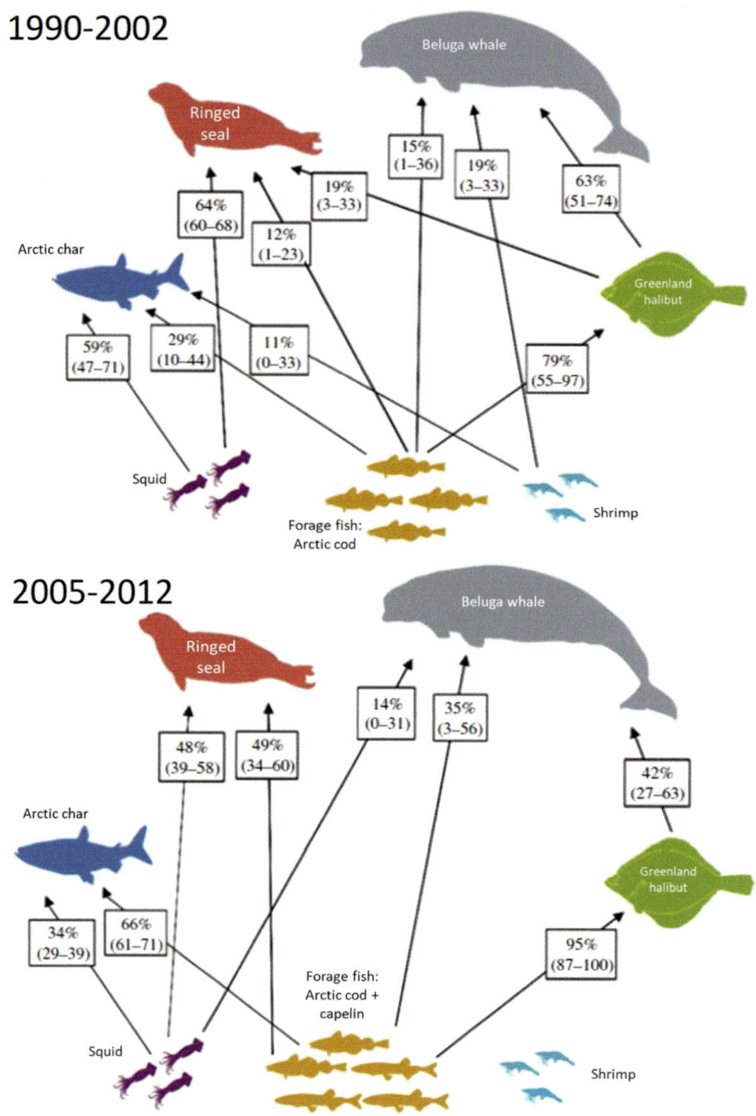

Figure 9.7a Results from stable isotope mixing models with median contributions (where over 10%) of prey species. Figures in brackets show 95% Bayesian credibility intervals. (From Yurkowski et al., 2018. Reproduced under the Creative Commons CC BY 4.0 Attribution licence, copyright 2018 Yurkowski et al.)

Question 9.7.1: How did the four top predators (Arctic char, ringed seal, beluga whale and Greenland halibut) respond to changes in forage fish?

A. It made no difference.
B. They all increased the proportion of forage fish in their diet.

C. Their diets switched from predominantly capelin to mostly Arctic cod.
D. Each species responded differently with some increases and some decreases in feeding on forage fish.

Question 9.7.2: What do these findings suggest about these predators? (No more than 20 words are needed.)

Question 9.7.3: What do these findings suggest about forage fish population sizes and composition in Cumberland Sound? (No more than 40 words are needed.)

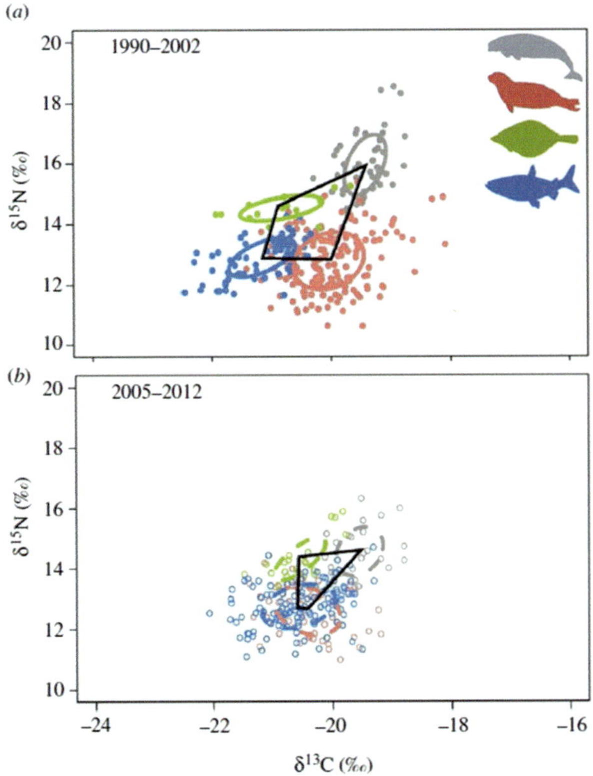

Figure 9.7b $\delta^{13}C$ and $\delta^{15}N$ for samples from beluga whale, Arctic char, ringed seal and Greenland halibut. Coloured ellipses show the 40% isotopic niche size of each predator; black lines join the centroid for each species to show community metric total area. Colour coding as for the previous figure. (From Yurkowski et al., 2018. Reproduced under the Creative Commons CC BY 4.0 Attribution licence, copyright 2018 Yurkowski et al.)

Question 9.7.4: How have the isotopic niches of the four predatory species changed between the two time periods? See Figure 9.7b.

A. The predators ate fewer prey from 2005 to 2012.
B. The changes in isotopic niche were primarily driven by Greenland halibut and Arctic char shifting their dietary preference from invertebrates to forage fish.
C. The predators ate a greater diversity of prey from 2005 to 2012.
D. The predators ate a reduced diversity of prey from 2005 to 2012.

Question 9.7.5: What are the implications for interactions between these predators as climates warm? (No more than 20 words are needed.)

Exercise 9.8: Impact of warming environments on limpets

The impact of climate change on key species in communities can buffer or exacerbate the impacts of warming. Researchers ran experiments to assess the impact of warming on limpets (*Lottia* spp.), an important coastal herbivore and thus the wider community, at Salt Spring Island, British Columbia, Canada. They used 'settlement' boards that were black (warming treatment) or white (ambient treatment) and copper sheeting to exclude limpets from half of each board. Any limpets that were able to settle on the exclusion boards were removed manually (Figure 9.8a).

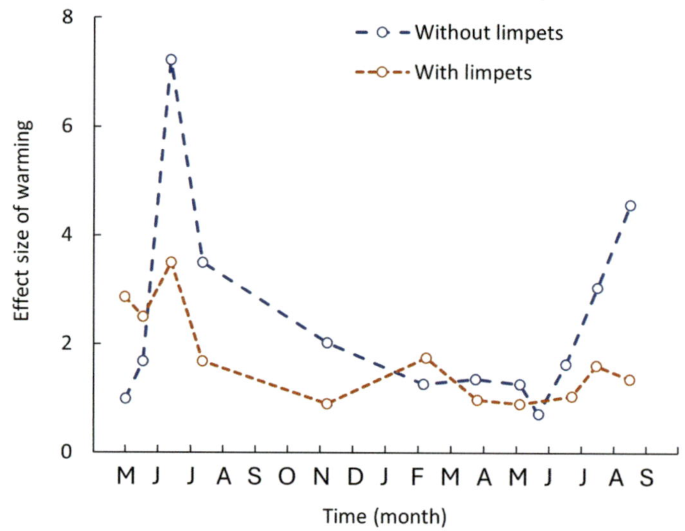

Figure 9.8a Effect size of warming on the composition of the communities establishing on the boards over the duration of the experiment (18 months). (Adapted from Kordas et al., 2017. Reproduced under the Creative Commons Attribution licence 4.0, copyright 2017 Kordas et al.)

Question 9.8.1: Overall, what impact did limpets have on the community composition?

A. No effect.
B. The presence of limpets throughout the experiment reduced the impact of warming.
C. Where limpets were excluded, communities took longer to recover following summer high temperatures.
D. Where limpets were present, communities took longer to recover following summer high temperatures.

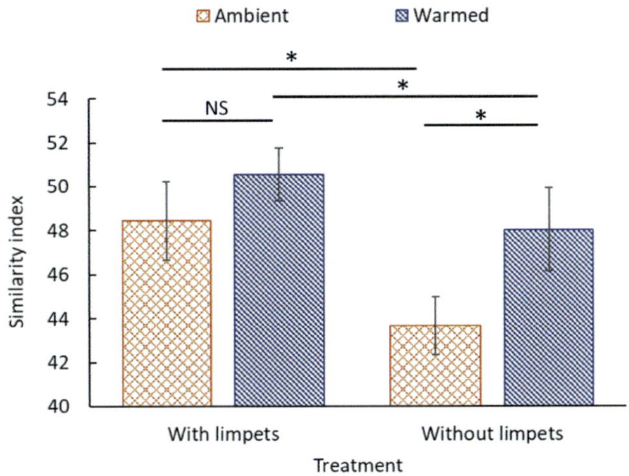

Figure 9.8b Bar chart showing Bray–Curtis similarity index of cover of all taxa when limpets were present or absent and in warmed or ambient temperature treatments. Bars are mean ± 1 SE. (Adapted from Kordas et al., 2017. Reproduced under the Creative Commons Attribution licence 4.0, copyright 2017 Kordas et al.)

Question 9.8.2: What effect do limpets have on the similarity of communities experiencing warming? See Figure 9.8b. (No more than 20 words needed.)

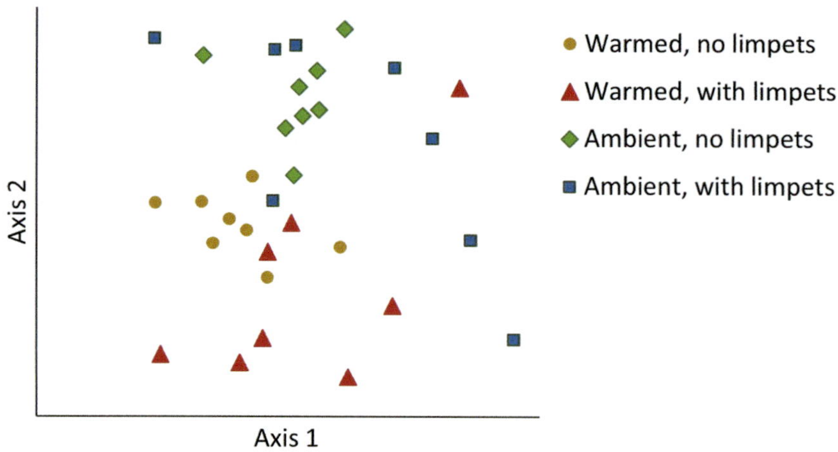

Figure 9.8c Non-metric multidimensional scaling analysis for variation in warming and herbivory treatments. Points represent the successional trajectory for a single plate. Stress: 0.22.

Question 9.8.3: Using Figure 9.8c, select the correct statements from the following list. (More than one answer is correct.)

A. The successional trajectory of warming treatments without limpets showed greater diversity than warming treatments with limpets.
B. The successional trajectory of warming treatments without limpets showed relatively low diversity.
C. The successional trajectory of ambient treatments with limpets showed the greatest diversity.
D. The successional trajectory of ambient treatments without limpets was more diverse than the ambient treatment with limpets.

Question 9.8.4: The researchers noted differences in timing of recruitment of organisms in these coastal communities: spores of large algae settled in available gaps in January and barnacles settled in available gaps in April. They also noted high variability in numbers of limpets in both time and space. How might these observations help to explain the findings? (Aim for around 100 words for a full explanation.)

Question 9.8.5: In no more than 20 words, set out the implications of these experimental findings.

10

References, answers and explanations

Where text answers include information in square brackets, this represents additional explanation or details not expected to be given in the answer.

Most questions have a detailed answer to explain the reasoning. Where a question is more straightforward (e.g. reading something directly from a graph with limited interpretation) no additional explanation is given.

Preface

References

Brehaut, L. and Brown, C.D., 2022. Wildfires did not ignite boreal forest range expansion into tundra ecosystems in subarctic Yukon, Canada. *Plant Ecology*, **223**(7), pp. 829–847.

Fordyce, A.J., Ainsworth, T.D. and Leggat, W., 2021. Light capture, skeletal morphology, and the biomass of corals' boring endoliths. *mSphere*, **6**(1), p. e00060-21. doi: 10.1128/msphere.00060–21.

Greenacre, M., Groenen, P.J.F., Hastie, T., D'enza, A.I., Markos, A. and Tuzhilina, E., 2022. Principal component analysis. *Nature Reviews Methods Primers*, **2**(1), p. 100.

DOI: 10.1201/9781003644095-10

Lin, Y., Wu, H., Liu, D., Li, Y., Kang, Y., Zhang, Z. and Wang, W., 2023. Patterns and drivers of soil surface-dwelling Oribatida diversity along an altitudinal gradient on the Changbai Mountain, China. *Ecology and Evolution,* **13**(5), p. e10105.

Schneider, M., 2020. Error bars – frequently used but rarely questioned. Data Analysis Tools. Available at: https://dataanalysistools.de/2020/03/29/error-bars-frequently-used-but-rarely-questioned/ (accessed 5 August 2025).

Chapter 1

Exercise 1.1

Bai, Y., Han, X., Wu, J., Chen, Z. and Li, L., 2004. Ecosystem stability and compensatory effects in the Inner Mongolia grassland. *Nature,* **431**(7005), pp. 181–184.

Question 1.1.1: A, B, C, D

If complementarity is occurring, we would expect functional groups to show negative correlations—as one group decreases, another increases. Positive correlations show that both functional groups are responding in the same way, both increasing or decreasing.

Question 1.1.2: C

The level which is most stable will show the least variation, and this occurs at the community level.

Exercise 1.2

Naeem, S., Thompson, L.J., Lawler, S.P., Lawton, J.H. and Woodfin, R.M., 1994. Declining biodiversity can alter the performance of ecosystems. *Nature,* **368**(6473), pp. 734–737.

Question 1.2.1: A

Transmittance is correlated with the inverse of productivity, so the lowest transmittance (high diversity) has the highest productivity.

Question 1.2.2: C, F

Make sure that you have read the explanatory information and interpreted the graphs correctly. All the treatments were carbon sinks, but the high diversity treatment accumulated the most carbon, as shown by more negative C fluxes.

Question 1.2.3

The more complex vegetation structure in the high diversity treatment filled more of the available space in the ecotrons and thus intercepted more light, had a higher rate of photosynthesis and retained more C within the system.

Exercise 1.3

Maron, J. and Marler, M., 2007. Native plant diversity resists invasion at both low and high resource levels. *Ecology*, **88**(10), pp. 2651–2661.

Question 1.3.1: C

The causal relationship is more likely to be that it is harder for knapweed to establish in species-rich communities than the other way round (A, B). Graph C suggests that knapweed can establish in communities regardless of species richness (D), but it is less able to persist in more diverse communities (graphs A and B).

Question 1.3.2

The species richness of the community does not affect the germination of knapweed seedlings. Presumably, they are able to establish when other species in the community are dormant. However, species-rich communities contain fewer mature knapweed plants, so diversity may affect the ability of knapweed to persist and become dominant.

Exercise 1.4

Wiser, S.K., Allen, R.B., Clinton, P.W. and Platt, K.H., 1998. Community structure and forest invasion by an exotic herb over 23 years. *Ecology*, **79**(6), pp. 2071–2081.

Question 1.4.1: C

Question 1.4.2

Dominant species become dominant by being better at acquiring resources and outcompeting other species. It may be more difficult for an invading species to establish itself in plots with dominant species.

Exercise 1.5

Becker, P., 1992. Colonisation of islands by carnivorous and herbivorous heteroptera and coleoptera: Effects of island area, plant species richness, and 'extinction' rates. *Journal of Biogeography*, **19**(2), pp. 163–171.

Question 1.5.1: D

Equilibrium theory predicts that small islands far from sources of immigrants will have low species diversity, and large islands close to sources of immigrants will have high species diversity. It doesn't make predictions about relationships between numbers of plants and animals (A). Despite the prediction of the theory, in this example we don't have any data on numbers of plant species in relation to island area (B); we also don't have any data on distance from the mainland in relation to numbers of beetles (C).

Question 1.5.2: A

More habitats will support a greater number of species, and larger islands will tend to contain a wider range of habitats (but this need not always be the case).

Here, it looks as if the number of plant species is decoupled from the area of the island, i.e. the relationship between beetle number and island area isn't significant for beetles, but the relationship between plant species and beetles is significant (A). The size of species shouldn't make any difference to the general pattern of the species-area relationship (B). Option C might be true but won't explain the general patterns seen here; we don't have any data here in relation to distance from the mainland, nor is there any suggestion that beetles are less able to disperse. Flying beetles may be better able to disperse (D).

Exercise 1.6

Packer, A. and Clay, K., 2000. Soil pathogens and spatial patterns of seedling mortality in a temperate tree. *Nature*, **404**(6775), pp. 278–281.

Question 1.6.1: B

Check that you are looking at the correct axes for each set of data in Figure 1.6a. Triangles and diamonds (lines showing increases) are the probability of seedling survival, not the number of seedlings that germinated (A). The dashed green line with squares shows the decrease in number of seedlings that germinated further away from the parent tree (B). We can't tell if soil conditions influenced the findings in this experiment (C) as we don't know the number of seeds that arrived at different distances; it could be that more germinated close to the parent tree simply because more fell there. These data do support the Janzen–Connell hypothesis (D) as higher numbers of seeds germinated close to the parent, but they are more likely to survive farther from the parent tree.

Question 1.6.2: B

This is calculated from the number of seeds that germinated (approximately 250) multiplied by their probability of surviving to 16 months (approximately 0.6). This gives the answer 150.

Question 1.6.3: D

The first part of this exercise shows that whilst more seeds germinated closer to a parent tree, in the long run survival was less good close to a parent tree (A). We have no information about seed predation (B) or grazing (C) so we can't tell if these answers are correct. The presence of specialised soil fungi (D) seemed the most likely answer as sterilising the soil made a big difference to survival where seeds had arrived close to the parent tree. If this pattern was driven by competition with parents (E), it wouldn't make any difference to survival whether the soil was sterilised or not for the seedlings germinating close to their parent.

Exercise 1.7

Ogada, D.L., Gadd, M.E., Ostfeld, R.S., Young, T.P. and Keesing, F., 2008. Impacts of large herbivorous mammals on bird diversity and abundance in an African savanna. *Oecologia*, **156**(2), pp. 387–397.

Question 1.7.1: D

Question 1.7.2: D

Question 1.7.3: E

Question 1.7.4: E
This question checks that you understand how to interpret graphs with letters showing significant differences between treatments. The relationship between megaherbivore grazing and bird diversity was tested directly by having different treatments relating to grazing. However, the link between bird diversity and tree canopy area wasn't directly tested. It seems likely that tree canopy area affected bird diversity, but to be sure, areas without grazing would need to be manipulated to have different levels of canopy cover, and the bird diversity measured. Correlation isn't causation; you need to do more experiments to be sure.

Question 1.7.5
Grazing by herbivores and megaherbivores significantly reduced tree canopy area. The positive relationship between tree canopy area and bird diversity *suggests* that the reason for reduced bird diversity where there are megaherbivores *could* be due to reduced canopy area.

Exercise 1.8

Bobbink, R. and Willems, J.H., 1987. Increasing dominance of *Brachypodium pinnatum* (L.) beauv. in chalk grasslands: A threat to a species-rich ecosystem. *Biological Conservation*, **40**(4), pp. 301–314.

Question 1.8.1: B, C
Make sure you have read the correct axes for both of the lines; it's easy to get these muddled up!

Question 1.8.2
Without more information, we can't really assess whether the increase in heath false brome reduced the diversity of the grassland (e.g. through competitive exclusion) or if reduced diversity allowed heath false brome to take advantage of vacant niches. The more likely explanation is that grazing kept heath false brome in check, preventing it from outcompeting other species.

Chapter 2

Exercise 2.1

Flood, P.J., Loftus, W.F. and Trexler, J.C., 2024. Do community changes persist after irruptive population dynamics? A case study from an invasive species boom and bust. *Oecologia*, **205**(3), pp. 445–459.

Question 2.1.1: B

Question 2.1.2: B

Question 2.1.3: A, J

These details are shown on Figure 2.1a: k is the number of dimensions and the stress value is below 0.2, so is an acceptable (but not excellent) fit for the data. Bluefin killifish (LUCGOO) is shown next to the 'boom' phase and spotted sunfish (LEPPUN) is shown next to the 'bust' phase.

Question 2.1.4

The community became much more variable, shown by the much bigger ellipse in Figure 2.1a based on the mean and 95% confidence intervals for this time period. During this time, the community was associated with flag fish.

Question 2.1.5: False

There was overlap of some of the ellipses for 1978–1984, 1985–1988, 1991–1995 and 2000–2011, but there were differences between the time periods. Only the ellipses for 1978–1984 and 1985–1988 were for consecutive time periods (see Figure 2.1a). This suggests that community composition changed over time even when jewel fish were at low frequency.

Question 2.1.6: C, E, F, G, I

These five species were the ones which showed a significant difference in density between the boom-and-bust phases. It wasn't determined by the most or least abundant species but the one that showed the greatest change.

Question 2.1.7

Answer: The fish community appeared to be very similar during the 'boom' and 'bust' phases, suggesting little change.

(Reason 1): The community was variable even when jewel fish populations were low.

(Reason 2): The new community assemblage could be due to *hysteresis* (where a return to a previous community structure requires greater changes in the constituent populations than were needed to cause the change initially).

(Reason 3): Alternatively, high numbers of jewel fish may have reset priority effects (where species suppressed below a threshold level find it difficult to increase due to competition or predation).

Question 2.1.8

The larger spotted sunfish was less susceptible to predation by jewel fish. It is a generalist feeder, so might be less affected by competition with other fish. It was the only species that increased during the jewel fish 'boom' phase.

Exercise 2.2

Schwilk, D.W., Keeley, J.E. and Bond, W.J., 1997. The intermediate disturbance hypothesis does not explain fire and diversity pattern in fynbos. *Plant Ecology*, **132**(1), pp. 77–84.

Question 2.2.1: B, E, F

Diversity was highest on the site burned every 5 years, so A is incorrect but E is right. The error bars suggest that there was no significant difference in diversity between sites burned every 20 and 40 years, so B is right. The Intermediate Disturbance Hypothesis predicts that intermediate levels of disturbance will result in the highest biodiversity (too much disturbance prevents many species from persisting; too little allows stronger competitors to dominate and exclude other species) but this wasn't seen at this site, so C and D are both wrong but F is correct.

Question 2.2.2

Sites experiencing the intermediate burning regime still had the lowest species richness even when measured at 1000 m^2. However, sites with the 40-year burning regime had higher species richness when measured at 1000 m^2.

Question 2.2.3

Very frequent fires will prevent dominance by larger species. This leaves more space in which smaller plants can be 'packed'; more smaller plants are more likely to result in greater diversity.

Question 2.2.4: C

Conservationists need to consider whether promoting biodiversity is always the most important aim. It isn't possible to have the highest diversity and maximise proteas, as they need a lower level of disturbance unless sites are very large.

Exercise 2.3

Auld, T.D., 1987. Population dynamics of the shrub *Acacia suaveolens* (Sm.) Willd.: Survivorship throughout the life cycle, a synthesis. *Australian Journal of Ecology*, **12**(2), pp. 139–151.

Question 2.3.1: A, B, D, E

Answer A is correct because each graph has a portion on the left-hand side showing at least some seeds were killed. B is correct because the area showing the seeds killed was greatest in the high intensity fire diagram. C is incorrect because the area for seeds killed was smaller under moderate intensity fires. D and E are correct because where seed depth is 0, the percentage killed was 100% (this is true also for high-intensity fires).

Question 2.3.2: B

From the data shown here, we can't tell whether fire is *needed* for seed germination (A). The area shown for seed germination following high-intensity fires was greater than for low and moderate intensity (B); you need to judge this by eye as these data aren't shown. Option C initially looks as if it is the same answer as B, but it relates to *numbers* of seeds rather than percentages and we don't have those data here to be able to assess if it is correct. D and E are both incorrect as the areas shown in these blocks were lower than for high-intensity fires.

Question 2.3.3: C

Answer A is incorrect; although the area of dormant seeds was greatest under low-intensity fires, we don't have any information about what stimulates this dormancy (correlation isn't causation). Seeds need to be buried at least 4 cm deep to survive fires (C). Although 2 cm was sufficient under low-intensity fires, a small proportion was killed under moderate intensity and a greater proportion under high-intensity fires (B).

Question 2.3.4

The communities of sweet wattle would be able to survive this burning regime; only small numbers of seeds would be lost and over half of the seeds would germinate, as long as seeds were not buried over 2 cm deep. However, most seeds will remain dormant in the seed bank. These seeds may need hotter fires to stimulate germination, but we can't determine that without additional information. We also need to know for how long seeds can remain dormant in the seed bank.

Exercise 2.4

Álvarez-noriega, M., Baird, A.H., Dornelas, M., Madin, J.S. and Connolly, S.R., 2018. Negligible effect of competition on coral colony growth. *Ecology*, **99**(6), pp. 1347–1356.

Question 2.4.1

Trade-off 1: Species that grew faster and produced more offspring (*Acropora* corals) were less stable and thus more vulnerable during storms than massive corals (*Goniastrea* species).

Trade-off 2: Species which were better at overtopping other species (*Acropora* corals) did less well than massive corals (*Goniastrea* species) in direct contact competition.

Question 2.4.2

Regular disturbance will prevent overtopping species from becoming competitively dominant and excluding other species. The ability to grow fast allows overtopping species to recover more rapidly following disturbance. These interactions will promote high species diversity in coral communities.

Exercise 2.5

Adam, T.C., Brooks, A.J., Holbrook, S.J., Schmitt, R.J., Washburn, L. and Bernardi, G., 2014. How will coral reef fish communities respond to climate-driven disturbances? Insight from landscape-scale perturbations. *Oecologia*, **176**(1), pp. 285–296.

Question 2.5.1

The starfish killed the coral, but the some of the dead skeletons remained, retaining structural complexity, which may be important for recovery of the reef. The hurricane destroyed live and dead coral.

Question 2.5.2: C, D, E

Whilst option A is true, it doesn't actually answer this question. It's unlikely that option B is correct as roving herbivores were not going to be competing for the same resources as coral-dwelling species. C, D and E are all more likely to be the case—fish which spend their larval stage in more sheltered areas will be able to produce a supply of new individuals able to take advantage of gaps following disturbance.

Question 2.5.3

As roving herbivores increased following the outbreak of crown of thorns starfish and cyclone Oli, it is likely that algae continued to be suppressed and wouldn't prevent coral from recovering in the longer run.

Exercise 2.6

Daleo, P., Alberti, J., Pascual, J., Canepuccia, A. and Iribarne, O., 2014. Herbivory affects salt marsh succession dynamics by suppressing the recovery of dominant species. *Oecologia*, **175**(1), pp. 335–343.

Connell, J.H. and Slatyer, R.O., 1977. Mechanisms of succession in natural communities and their role in community stability and organisation. *The American Naturalist*, **111**(982), pp. 1119–1144.

Question 2.6.1: A

There were no treatments in the marsh matrix, so grazing and clonal spread occurred as they would normally.

Question 2.6.2

The controls where clonal spread and grazing were allowed were also cleared at the start of the experiment. Clonal spread and grazing occurred in the marsh matrix, suggesting that recovery from bare ground takes longer than 6 years.

Question 2.6.3: B

Pickleweed facilitated the clonal spread of denseflower cordgrass (the dominant salt marsh species in the long run) but only in the absence of grazing.

Question 2.6.4: A, D

In treatments with grazing, seashore saltgrass and denseflower cordgrass both had lower biomass than when grazers were excluded.

Question 2.6.5: B

Denseflower cordgrass is a late successional species, and herbivores can inhibit its establishment by eating it. Seashore saltgrass is a pioneer species that facilitates the growth of denseflower cordgrass, so the reduction of this species through grazing also inhibited succession.

Exercise 2.7

Menge, B.A., Bracken, M.E.S., Lubchenco, J. and Leslie, H.M., 2017. Alternative state? Experimentally induced *Fucus* canopy persists 38 yr in an *Ascophyllum*-dominated community. *Ecosphere*, **8**(3), pp. e01725.

Question 2.7.1: B

In the 2010s, the upper shore was more heterogeneous than the lower shore as shown by the much larger ellipse on the upper shore graph, so B is correct and A is the wrong way around. C is wrong because the difference that drove the pattern shown here is from reduced diversity in the 2010s between the upper and lower shore. The results from the 1970s on the upper shore were contained within the results ellipse for the 2010s, so there isn't really any difference between the samples taken at the two time points on the upper shore.

Question 2.7.2: A

These plots had no species removed from them; the knotted wrack and barnacles were left as they were on these plots.

Question 2.7.3: False

The control (see previous question) where no species were removed also showed a large reduction in knotted wrack, so the lower percentage cover of this species in the other two treatments couldn't be solely due to slow recovery. It suggests a wider change had occurred that reduced the cover of this species more generally.

Question 2.7.4

There's no sign that barnacles facilitated establishment of either species—there was no difference in the percentage cover of brown algae with or without barnacles, nor any difference for knotted wrack with and without barnacles.

Question 2.7.5

Although brown algae had a higher percent cover where knotted wrack was removed, it did not become competitively dominant. It could be that it was unable to colonise below the knotted wrack canopy.

Question 2.7.6: Alternative stable state

After such a long time without recovering to its former cover, it could be that this ecosystem has 'flipped' to a different state. This is a difficult question to answer, so don't worry if you didn't get it!

Exercise 2.8

Pearson, D.E., Ortega, Y.K. and Lekberg, Y. 2024. Invaders break assembly rules to beat the natives: How cheatgrass cheats. *NeoBiota*, **96**, pp. 299–324.

Question 2.8.1: B, D

Although the bar for naturalised species was the highest here (A), the shared letters show there was no difference between establishment in native and naturalised species starting communities (B). You need to take care with this graph to be sure you know what it is telling us. It isn't about the numbers of plants established in each starting community, so C is wrong. We don't have any data about numbers of individual plants in each starting community, only how many perennials were able to establish into each community. D is correct; fewer perennials established in the invasive annual starting communities. E is incorrect; we don't have any information here on numbers of invasive species used in these plots.

Question 2.8.2: Facilitation

The presence of annuals appeared to help the perennials establish. The control plots had no starting community, and the perennials established less well in those plots.

Question 2.8.3: Inhibition

The invasive annuals seemed to make it more difficult for perennials to establish.

Question 2.8.4

Native annual and naturalised species were initially present at high cover, but invasive annuals had a higher percentage cover. After 2 years, the cover of native and naturalised annuals had reduced considerably, but the cover of invasive annuals was still very high. After 2 years, additional species had started to colonise the native and naturalised annual communities; this was likely due to the facilitating effect of the native annuals. However, there was very little colonisation of the invasive annual communities, suggesting that these species inhibited succession to communities of perennial species.

Chapter 3

Exercise 3.1

Davis, J.M., Rosemond, A.D., Eggert, S.L., Cross, W.F. and Wallace, J.B., 2010. Long-term nutrient enrichment decouples predator and prey production. *Proceedings of the National Academy of Sciences*, **107**(1), pp. 121–126.

Question 3.1.1: A

They expected to see a steady increase in primary consumer production after year 2 when the nutrient additions started, with a corresponding increase in predator production in the treatment data; controls should have remained similar to their starting level (A). B shows a steady increase from the start of the experiment, even before nutrients were added. C shows the increases starting from before the nutrient additions, and there is no difference between the treatment and controls. D shows increases following nutrient additions, but there is no difference between the treatment and controls.

Question 3.1.2: B

Question 3.1.3: F

Question 3.1.4: D

The stone cases of the caddis flies (A) wouldn't be included in the biomass measurements. *Pycnopsyche* feed on detritus and are described as part of the primary consumer community, so wouldn't be competing with predators (B). There has been no previous mention of fish, so this is a complete distraction (C). If something like this had happened in the experiment, it would have been included in the explanatory information. Answer D is correct—the stone cases give this group a certain amount of protection from predators.

Question 3.1.5

The aquatic community became dominated by large, well-protected invertebrates that could take advantage of the increased nutrients. The mouths of gape-limited predators were too small for them to prey on these large primary consumers, so they were unable to benefit from increased nutrients in the system.

Question 3.1.6

Some energy will be transported downstream and consumed by predators. Energy may also be transferred to terrestrial systems via predators that feed on emerging adult insects such as caddis flies.

Exercise 3.2

Vitousek, P.M. and Farrington, H., 1997. Nutrient limitation and soil development: Experimental test of a biogeochemical theory. *Biogeochemistry*, **37**(1), pp. 63–75.

Question 3.2.1

N and P commonly limit plant growth. The factorial design of the experiment enabled them to separate out the effect of each nutrient.

Question 3.2.2: B

Question 3.2.3: A

Question 3.2.4: B, C, D, F

Watch out here for the overlapping error bars.

Question 3.2.5

N was limiting on the 300-year-old island, most likely because leguminous plants had not established sufficiently to add this nutrient into the soil; however, young volcanic soils were rich in P compounds. P was limiting on the 4-million-year-old island because this nutrient had now weathered out of the volcanic soils, but N had built up due to the long establishment of legumes. N and P were both limiting on the 20,000-year-old island—legumes had not established enough to increase the availability of N, but weathering of P had reduced its prevalence.

Exercise 3.3

Wild, C., Huettel, M. and Tollrian, R., 2004. Rapid recycling of coral mass-spawning products in permeable reef sediments. *Marine Ecology Progress Series*, **271**, pp. 159–166.

Question 3.3.1: B

The graph shows O_2 consumption (i.e. a reduction in O_2) in the sediment, rather than an increase, so A and D are incorrect; phytoplankton may live on the surface of the sediment and were probably nutrient limited. An increase in population size and thus photosynthesis would produce O_2 during the day and CO_2 during the night, but this doesn't really answer the question (C). A decrease in O_2 shows it was being used for respiration by the benthic community (B).

Question 3.3.2

The sediments were probably low in nutrients and the productivity of the organisms there was nutrient limited.

Question 3.3.3

Many unfertilised eggs dropped straight to the sediment, but fish and other organisms fed on the dead eggs, slowing their journey to the sediment. Some of the nutrients were retained within the organisms' bodies, allowing them to grow or reproduce, but much was returned to the system within a few days as faeces.

Exercise 3.4

Hanlon, R.D.G. and Anderson, J.M., 1980. Influence of macroarthropod feeding activities on microflora in decomposing oak leaves. *Soil Biology and Biochemistry*, **12**(3), pp. 255–261.

Question 3.4.1: C

The presence of six woodlice reduced fungal biomass more than the presence of two woodlice, so the most likely explanation is that they ate the fungi as well as the bacteria.

Question 3.4.2

Woodlice consumed leaf litter (and fungi) but they couldn't digest all the plant material. While moving through the digestive systems of the woodlice, this material was 'processed' and broken up into smaller parcels of plant cells. These faeces provided a considerably larger surface area that was colonised by bacteria. More woodlice resulted in more processing and so more bacteria were present.

Exercise 3.5

Shigyo, N., Umeki, K. and Hirao, T., 2024. Soil microbial identity explains home-field advantage for litter decomposition. *New Phytologist*, **243**(6), pp. 2146–2156.

Question 3.5.1: A, B, G, H

If the home-field advantage was more important, litter from the upland species, hemlock, would decompose faster in upland soil, regardless of site (A, B). Litter from beech, the lowland species, would decompose faster in lowland soil, regardless of site (G, H).

Question 3.5.2: A, C, F, H

If the environment was the main driver of decomposition, litter from hemlock would decompose faster at upland sites, regardless of the type of soil in which the litter was buried (A, C). Likewise, beech would decompose faster at the lowland site (F and H).

Question 3.5.3: A

If you got the answers to the previous questions the right way round, you should be able to see from the graphs that this study shows home-field advantage was the more important determinant of litter decomposition.

Exercise 3.6

Buitenwerf, R., Stevens, N., Gosling, C.M., Anderson, T.M. and Olff, H., 2011. Interactions between large herbivores and litter removal by termites across a rainfall gradient in a South African savanna. *Journal of Tropical Ecology*, **27**(4), pp. 375–382.

Question 3.6.1

Phenolics and lignin affected palatability of litter, making it harder for the fungi that termites farm to digest the plant material. Although the litter in the bags was high quality, there may be fewer termites visiting areas with poor-quality plant material.

Question 3.6.2: D, E

Termites may prefer short vegetation (A), but the situation here seems more complicated as the wetness of the site also affected the amount of litter taken by termites. If either B or C were correct, this would suggest poor experimental design. By default we are left with D and E: termites did prefer to pick up material dropped by larger grazers, and there would be considerable nutrition left in the dung of these grazers which was more accessible to decomposers after having been through the digestive system of another animal.

Question 3.6.3

On wet soils, termites may compete with large herbivores. On dry soils, the relationship is commensal, where one species benefits but the effect is neutral for the other species. [This is determined by comparing litter removed in dry areas, where there is an increased removal in grazed plots and litter removed in wet areas, where less litter is removed in grazed plots]

Exercise 3.7

Gilbert, L., Jeanniard-du-Dot, T., Authier, M., Chouvelon, T. and Spitz, J., 2023. Composition of cetacean communities worldwide shapes their contribution to ocean nutrient cycling. *Nature Communications*, **14**(1), p. 5823.

Question 3.7.1

A = baleen whales (the eigenvalue for N runs through the middle of the ellipse for A).

B = deep divers (the eigenvalue for Cu runs through the middle of the ellipse for B).

C = small cetaceans (the eigenvalue for P runs through the middle of the ellipse for C).

Question 3.7.2: D

Add together the percentages for each axis (62.8%). There are as many dimensions as there are variables, but these explain progressively less of the variation in the data. All the dimensions added together will explain 100% of the variation in the data.

Question 3.7.3

(i) Baleen whales feed 'lower down' the food chain on phytoplankton and zooplankton; other cetaceans fed on fish and squid so nutrients have passed through a number of trophic levels. (ii) Even if nutrients were low, their faster turnover and thus availability led to higher primary productivity.

Question 3.7.4: D

Baleen whales do not feed much in tropical waters (A), but this doesn't answer the question in relation to nutrient cycling. B is unlikely to be true as these groups forage in different strata of the ocean (and it doesn't answer the question in relation to nutrient cycling). C is incorrect as deep-diving cetaceans are bringing in a source of nutrients from deeper parts of the ocean, so D is correct.

Question 3.7.5: A, B

Question 3.7.6: C

Question 3.7.7: C

Exercise 3.8

Appoo, J., Bunbury, N., Jaquemet, S. and Graham, N.A.J., 2024. Seabird nutrient subsidies enrich mangrove ecosystems and are exported to nearby coastal habitats. *iScience*, **27**(4), p. 109404.

Question 3.8.1: B

$\delta^{15}N$ of atmospheric N is 0 ‰, so positive values show enrichment and negative values show depletion. The values in the table were positive, so showed enrichment (B). We haven't been given any information on significant differences between the values, but these can be calculated using the standard deviations shown and the sample size. Divide the SD by the square root of N to derive the SE; multiply the SE by 1.96 to give the 95% confidence interval. We can see that these definitely overlapped for both N content of guano and $\delta^{15}N$ of guano, so there was no difference between the bird species for either of these factors. For more details on how to interpret error bars, see the Preface.

Question 3.8.2

Sediment, mangrove, gastropod, sesarmid crab, portunid crab.

The sesarmid crab and gastropod are both considered to be mid-trophic, but the values without seabirds suggest that sesarmid crabs were slightly higher up the food chain than gastropods.

Question 3.8.3

Seabirds brought nutrients into this ecosystem from the ocean areas where they fed on fish. This provided extra nutrients for mangrove growth and the associated food web.

Question 3.8.4: B

Carbon is rarely (if ever) limiting to plant growth. The N:P ratio suggests that P wasn't limiting, so by a process of elimination, it is most likely that N was limiting here.

Question 3.8.5

Despite the addition of nutrients at the seabird sites, these communities may still be nutrient limited for N at least.

Chapter 4

Exercise 4.1

Chesney, E.J. and Tenore, K.R., 1985. Oscillations of laboratory populations of the polychaete *Capitella capitata* (Type I): Their cause and implications for natural populations. *Marine Ecology Progress Series*, **20**(3), pp. 289–296.

Additional reference:

Keerthika, K., Padmavathy, P., Rani, V., Jeyashakila, R., Aanand, S. and Kutty, R., 2024. Evidence of microplastics in the polychaete worm (capitellids—*Capitella capitata*) (Fabricicus, 1780) along Thoothukudi region. *Environmental Monitoring and Assessment*, **196**(6), pp. 556.

Question 4.1.1: B

Exponential growth occurs when populations keep increasing. It is common as a population is growing from a small size, before conditions become limiting. Logistic growth occurs when a population becomes limited by a resource and remains at a more or less constant size. This population did not show logistic growth at any point in the growth cycle shown here.

Question 4.1.2

The population cycles shown here were probably caused by a time lag in the response of individuals to the increasing population size. The most likely time lag here was that between an adult female producing eggs and the offspring being recruited to the adult population and competing for food resources.

Question 4.1.3

Reduction in breeding led to almost complete cessation in larval settlement as the population reduced in size.

Exercise 4.2

Castagneyrol, B., Van Halder, I., Kadiri, Y., Schille, L. and Jactel, H., 2021. Host-mediated, cross-generational intraspecific competition in an herbivore species. *Peer Community Journal*, **1**, e61.

Additional reference:

Gripenberg, S., Mayhew, P.J., Parnell, M. and Roslin, T., 2010. A meta-analysis of preference–performance relationships in phytophagous insects. *Ecology Letters*, **13**(3), pp. 383–393.

Question 4.2.1: C

The bars look very similar in height and remember that standard errors are smaller than confidence intervals. The results of the statistical test confirm that there was no difference between the levels of this treatment; the P-value was above the cut-off level for significance of 0.05.

Question 4.2.2

Females may recognise pheromones produced by females that previously laid eggs on plants; frass (droppings) of insects also indicates previous or ongoing grazing; some females can detect chemicals produced by plants in response to grazing. The results suggest that box tree moths do not avoid laying eggs on plants which have previously been grazed.

Question 4.2.3: D

Answers A, B and C may be correct, but they don't answer the question specifically in relation to female choice of host plant. Induced defences by plants following grazing reduce growth in some species of herbivores, so it is worthwhile for females to avoid these plants. This appeared to be less important in box tree moths.

Question 4.2.4: B

The fitted lines cross, showing that the response of larval weight to the number of egg clutches doesn't follow the same pattern on box plants that have or have not been previously grazed. On plants without previous grazing, as egg clutches increase, larval weight goes down, but on plants that have been grazed previously, larval weight increases as the number of egg clutches increases. The other options show varying degrees of difference between the fitted lines, but the lines are more or less parallel, showing that the factors are responding in the same way but just to greater or lesser amounts.

Question 4.2.5

The presence of egg clusters on box plants did not reduce subsequent egg deposition. However, previous grazing did affect future generations: larva and chrysalis weights were both reduced by previous herbivory. Intergenerational, intraspecific competition was occurring even though individuals from different generations would never encounter each other directly.

Exercise 4.3

Branch, G.M., 1975. Intraspecific competition in Petella cochlear Born. *Journal of Animal Ecology*, **44**(1), pp. 263–281.

Question 4.3.1: C

The high-density line was farther to the left and this shows that a lower proportion of the population had a large shell size. For example, 50% of the population at high density had shells of 20 mm but the corresponding shell sizes for medium- and low-density populations were ~40 mm and ~58 mm, respectively. None of the shells in the high-density population was larger than ~50 mm, whereas the low-density population had shell lengths >70 mm.

Question 4.3.2: A

Make sure you chose the correct line—the low-density population is the bottom line shown with triangles. The y-axis is the *cumulative* percentage, so you need to subtract the percentages that are smaller than 10 mm long from the percentages that are 15 mm long, which is approximately 1%. For medium-density populations, this percentage was around 4%, and for high-density populations it was around 22%.

Exercise 4.4

Pfennig, D.W., 1992. Polyphenism in spadefoot toad tadpoles as a locally adjusted evolutionarily stable strategy. *Evolution*, **46**(5), pp. 1408–1420.

Question 4.4.1: D

All the bars were significantly different from each other except for the carnivores in the two types of pool. These two bars therefore shared a letter, which was different from the letters above both the bars for the omnivorous tadpoles.

Question 4.4.2: E

The 'best' life history varied according to the environment, so the only correct statement relates to the combination of omnivory in short-term pools.

Question 4.4.3

In the long run, it was probably better to be a carnivore, but the large number of omnivores that survived when pools persisted can flood the gene pool with individuals that also favour this morph. Pools will have to persist for a certain proportion of the time for this strategy to persist; if pools persist too infrequently, carnivores will be favoured.

Question 4.4.4

There was probably greater competition for food resources in the 100% carnivore pools, encouraging a change to morphs that can take advantage of alternative food sources.

Exercise 4.5

Pettifor, R.A., Perrins, C.M. and McCleery, R.H., 2001. The individual optimization of fitness: variation in reproductive output, including clutch size, mean nestling mass and offspring recruitment, in manipulated broods of great tits *Parus major*. *Journal of Animal Ecology*, **70**(1), pp. 62–79.

Additional reference:

Monaghan, P. and Nager, R.G., 1997. Why don't birds lay more eggs? *Trends in Ecology & Evolution*, **12**(7), pp. 270–274.

Question 4.5.1: B

Adding eggs to the nests reduced (rather than increased) the numbers being recruited to the adult population, relative to the unmanipulated nests (A), presumably due to increased competition or inability of parent birds to provide enough food. Having fewer young to raise (when eggs were removed from nests) may reduce intraspecific competition and survival to recruitment (C), but this wasn't enough to offset raising fewer young overall.

Question 4.5.2

Young birds need to be able to survive long enough and be sufficiently healthy to breed; large clutch sizes may reduce these. Adult birds which invest too much in raising young in one year may die early or be unable to breed every year.

Exercise 4.6

Murton, R.K., Westwood, N.J. and Isaacson, A.J., 1974. A study of wood-pigeon shooting: the exploitation of a natural animal population. *Journal of Applied Ecology*, **11**(1), pp. 61–81.

Question 4.6.1: C, D, E

It was assumed that predation (in the form of culling) would control (and reduce) the population, but the pigeons were limited more by intraspecific competition. When pigeons were removed from the population, survivorship of the remaining birds would have increased either through increased longevity or from higher breeding success, or both. If any food resources were left unexploited due to the culling, it is likely that birds from surrounding areas (where they were also experiencing high intraspecific competition) would move in to exploit this. If the pigeons were being killed by disease (A), the level of crop damage and the population size would likely have decreased. If other species were moving in to exploit the unused resource (B), damage would have stayed the same, but the population size of pigeons would have decreased. There was no indication that the population size reduced at all, so answer F can't be correct. This is described as a compensatory reaction to reduced intraspecific competition.

Exercise 4.7

Ens, B.J., Kersten, M., Brenninkmeijer, A. and Hulscher, J.B., 1992. Territory quality, parental effort and reproductive success of oystercatchers (*Haematopus ostralegus*). *Journal of Animal Ecology*, **61**(3), pp. 703–715.

Question 4.7.1: A, C, D, E, F

The data suggest that salt marsh was a better environment for oystercatchers to raise chicks, but we have no information on what made it better, so answers G, H and I might be correct, but we can't tell without more data. If the salt marsh was a better environment, there was likely to be high intraspecific competition for territories (a cost), and these birds would need to spend time defending territories (E) rather than feeding chicks, so it is likely that stronger and more experienced oystercatchers would be better at this (D). If, despite spending more time defending territories, these birds still managed to raise more chicks, they may also be better at finding food for them (F). Answer A is correct (from graph), and answer C is correct as this information is given in the question.

Question 4.7.2

Despite feeding on richer territories and having consistently higher capture rates of food, this was insufficient to outweigh the disadvantage of being farther from the feeding grounds. The time away from nests, the energetic costs of flying between nesting and feeding territories, and the shorter time that feeding grounds are available all seemed to reduce breeding success.

Exercise 4.8

Beggs, R., Tulloch, A.I.T., Pierson, J., Blanchard, W., Crane, M. and Lindenmayer, D., 2019. Patch-scale culls of an overabundant bird defeated by immediate recolonization. *Ecological Applications*, **29**(3), p. e01846.

Question 4.8.1: No

There was no difference between control and treatment patches even immediately after the cull. There were fewer very high population counts, but average numbers remained similar throughout the experiment.

Question 4.8.2: B

If noisy miners had bred more due to reduced intraspecific competition (A), there would have been a time lag between the cull and numbers increasing. Whilst there did seem to be natural variation in the population over time (C), the control and cull areas followed the same pattern and there was generally little difference between these. The most likely explanation is that the populations were limited by intraspecific competition, so the cull created vacant territories, which were quickly colonised by birds from surrounding areas where competition was still intense.

Question 4.8.3
Populations of noisy miners would need to be culled at a landscape scale to prevent rapid recolonisation from adjacent areas. Alternatively, continuous culling in key locations would keep removing noisy miners arriving from elsewhere.

Chapter 5
Exercise 5.1
Byun, C., 2023. Competition between pink muhly grass and native plant species: Is it really a harmful invasive species? *Ecosphere*, **14**(7), p. e4561.

Question 5.1.1: No
Although the percentage cover of native species varied considerably between the years, there was no statistically significant difference in cover of native plants between pots with and without pink muhly grass.

Question 5.1.2
Plants were grown in greenhouses under controlled conditions, so more trials should be carried out under field conditions. Plants were assessed only in two-way combinations, so pink muhly grass should be grown in plots with a greater range of species. It would also be good to assess dispersal ability via vegetative and sexual reproduction, ability to colonise disturbed ground, and how climate change might affect its ability to spread and suppress other species.

Exercise 5.2
Duckworth, R.A., Hallinger, K.K., Hall, N. and Potticary, A.L., 2017. Switch to a novel breeding resource influences coexistence of two passerine birds. *Frontiers in Ecology and Evolution*, **5**, p. 72.

Question 5.2.1: Western bluebird
This species started from very small numbers and increased over time.

Question 5.2.2: Mountain bluebird
This species initially had a high population, but this declined over time as western bluebirds increased, suggesting that this species was less good at competing and should, therefore, be better at colonising.

Question 5.2.3: A, B, C, D
All of these are potential reasons for the pattern shown in the figure!

Question 5.2.4
Mountain bluebirds were less competitive, so male birds used more time bringing food to fledglings.

Exercise 5.3

Feniova, I., Dzialowski, A.R., Bednarska, A., Brzeziński, T., Zilitinkevicz, N. and Dawidowicz, P., 2025. Shifts in competition outcomes between two *Daphnia* species in response to algal phosphorus content. *Oecologia*, **207**(2), p. 32.

Additional useful references:

Elser, J.J., Sterner, R.W., Gorokhova, E., Fagan, W.F., Markow, T.A., Cotner, J.B., Harrison, J.F., Hobbie, S.E., Odell, G.M. and Weider, L.W., 2000. Biological stoichiometry from genes to ecosystems. *Ecology Letters*, **3**(6), pp. 540–550.

Elser, J.J., Dobberfuhl, D.R., Mackay, N.A. and Schampel, J.H., 1996. Organism size, life history, and N:P stoichiometry: Toward a unified view of cellular and ecosystem processes. *BioScience*, **46**(9), pp. 674–684.

Isanta-navarro, J., Prater, C., Peoples, L.M., Loladze, I., Phan, T., Jeyasingh, P.D., Church, M.J., Kuang, Y. and Elser, J.J., 2022. Revisiting the growth rate hypothesis: Towards a holistic stoichiometric understanding of growth. *Ecology Letters*, **25**(10), pp. 2324–2339.

Question 5.3.1: C

The strongly significant interaction term shows that the species were doing different things in the low and high P treatments. We can't tell from this result alone how each was responding (but we can tell this from the figures that are shown in the following question).

Question 5.3.2: No, the reverse occurred.

Question 5.3.3

By definition, good colonisers can arrive and establish in new habitats before other species. *Daphnia magna* had a high growth rate and in the absence of competition took advantage of high P concentrations to grow fast and outcompeted species that arrive later; this is known as the priority effect. The P concentrations in the experiment may not have been high enough for this phenomenon to occur. The researchers also set up the cultures with equal starting biomass of each species, so this may have prevented *D. magna* from becoming dominant and excluding *D. longispina*.

Question 5.3.4

Small organisms have a higher ratio of surface area to volume, so lose a greater proportion of P when moulting compared with larger species.

Exercise 5.4

Sorvari, J., Huhta, E. and Hakkarainen, H., 2024. Coexistence of territorial competitor ants in fragmented boreal forest landscape. *Oecologia*, **206**(1), pp. 187–197.

Question 5.4.1: B, D

The figure shows the numbers of patches occupied by the ant species. There were 61 patches with *F. aquilonia* but only 17 patches with *F. polyctena*, suggesting that the former is more common (B). The chi square test is highly significant, showing that the distribution of the two species wasn't random and thus that something was affecting where the species were found (A) and it could well have been the other species of ant. C is speculation; we have no information about how or why this pattern of distribution has occurred. D is correct because although the model suggested a small number of patches should contain both species, in fact, none was observed.

Question 5.4.2: D

Niche partitioning (A) occurs when species occupy the same space but use resources differently; thus, they do not compete. Character displacement (B) occurs when a morphological feature of a species changes in relation to the presence of a competing species. Competitive release (C) is seen when a species expands its niche or area occupied in the absence of a competitor. Competitive exclusion (D) is the most likely explanation here, as the two species do not coexist.

Question 5.4.3

Formica aquilonia was the more abundant species and could be considered competitively superior. However, if *F. polyctena* was better at dispersing to unoccupied forest patches, it could stay 'one step ahead' of *F. aquilonia*. Once *F. aquilonia* had arrived at a patch, it would probably exclude any *F. polyctena* present. [The dispersing species are known as 'fugitives' and the interaction is called 'fugitive coexistence.']

Exercise 5.5

Pratte, I., Robertson, G.J. and Mallory, M.L., 2017. Four sympatrically nesting auks show clear resource segregation in their foraging environment. *Marine Ecology Progress Series*, (572), pp. 243–254.

Question 5.5.1: A

See answer for Question 5.4.2 for explanations of these terms.

Question 5.5.2

Common murre and razorbill fed in the same locations (inshore and benthic areas) and at a similarly higher trophic level, but common murre ate considerably more daubed shanny and fewer sand lance than razorbill, so they were making use of different prey species. Thick-billed murre also fed on daubed shanny but foraged in different locations (offshore and pelagic areas).

Exercise 5.6

Dubowy, P.J., 1988. Waterfowl communities and seasonal environments: Temporal variability in interspecific competition. *Ecology*, **69**(5), pp. 1439–1453.

Question 5.6.1: C

This line is plotted from the x,y points 0,1 to 1,0. Thus, points which fall in the upper triangle will tend to have higher y-values (habitat overlap) and higher x-values (foraging overlap). The converse is true for the lower triangle. As the foraging and habitat overlaps are expressed from 0 (no overlap) to 1 (complete overlap), the lower triangle represents reduced overlap in feeding and habitat use (C). The graphs don't give data on the association between food choice and habitat choice, so answers A and B can't be right. D is the opposite of C. The line in itself can't tell us whether or not competition is occurring, so E and F are also incorrect.

Question 5.6.2

In winter, there was less overlap between the food requirements and habitat use made by the ducks. In summer, there was much more overlap in resource requirements. This could be because food resources were more plentiful in summer, so there was reduced competition for food.

Exercise 5.7

These are made-up data to illustrate how a stable coexistence might look in an experiment. Useful resources on this topic include:

Begon, M. and Townsend, C.R., 2021. *Ecology: From Individuals to Ecosystems.* 5th edn. Chichester: Wiley.

Singer, F.D., 2024. *Ecology in Action.* 2nd edn. Cambridge: Cambridge University Press.

Question 5.7.1: C

Even though each trial had very different population sizes at the start of the experiment, all trials moved towards a similar combined population size of around 100 sp1 ml^{-1} and 300 sp2 ml^{-1}. This suggests that the most likely outcome was stable coexistence. If the two species can only coexist in the short term, the size of the starting populations is often a key factor in determining which species is successful and excludes the other.

Exercise 5.8

Harrington, L.A., Harrington, A.L., Yamaguchi, N., Thom, M.D., Ferreras, P., Windham, T.R. and Macdonald, D.W., 2009. The impact of native competitors on an alien invasive: Temporal niche shifts to avoid interspecific aggression. *Ecology*, **90**(5), pp. 1207–1216.

Question 5.8.1

The larger otter was able to out-compete smaller mink for food such as fish.

Question 5.8.2: Resource (or niche) partitioning

Question 5.8.3: C

There is no data provided to suggest that fish were in limiting supply, so answer A is less likely to be correct. B may be correct, but this is quite a complicated (and speculative) answer and there was no supporting evidence. C was the most likely reason—larger otters would probably come off better out of an interaction with smaller mink. D is what might happen once mink had changed their daily routine, but this question asks about the reason that they changed the foraging pattern.

Question 5.8.4: Temporal niche partitioning

[Niche or resource partitioning is partly right, but it's the fact that they have changed the time of the day that they are active that is particularly striking here.]

Question 5.8.5: D

Answer A is not true; the mammals on which mink preyed were found in terrestrial areas (and probably in greater numbers). There were no data to support B that birds are a better food (they may actually be more difficult to catch). Although mink have had a severe impact on water voles in the UK, this species only ever provided a small proportion of the mink's diet; if an animal were that dependent on one type of prey species, it's unlikely it would have done so well as an invasive species (C). By a process of elimination, D is the most likely answer and makes more sense in relation to favoured prey items.

Question 5.8.6: No, it seems unlikely.

Chapter 6

Exercise 6.1

Prior, K.M. and Palmer, T.M., 2018. Economy of scale; third partner strengthens a keystone ant-plant mutualism. *Ecology*, **99**(2), pp. 335–346.

Question 6.1.1: B

Both species gained: the thorn tree was protected against grazing elephants (but not other types of grazer), and the ants gained food from nectar.

Question 6.1.2: B

Again, both species benefited: the scale insects were protected from predators by the ants, and the ants gained additional food (honeydew).

Question 6.1.3: A and/or B

This is trickier to answer with only these data. Scale insects are herbivores that take phloem from the tree, so the best description would be parasitic. However, the graphs show that when the scale insects were not present, the tree suffered substantially more grazing damage from elephants, so there was potentially an indirect gain from having scale insects.

Question 6.1.4

In the absence of scale insects and ants, trees may experience more grazing damage from elephants, and there could be a shift from a woodland habitat to a more open savannah.

Question 6.1.5: A

Given the level of damage inflicted on trees, it is unlikely that bad-tasting scale insects would make any difference (B). The information in this question tells us that the ants tended the scale insects rather than eating them, so C isn't correct. There is nothing to suggest that the scale insects induced growth of thorns (D), so that leaves A as the most likely answer.

Question 6.1.6

The most likely reason that the presence or absence of scale insects only affected grazing by elephants is that other grazers/browsers were not deterred by ants, so their level of activity made no difference. Vertebrate browsers may also be deterred to some extent by the trees' defensive thorns.

Exercise 6.2

Silliman, B.R. and Newell, S.Y., 2003. Fungal farming in a snail. *Proceedings of the National Academy of Sciences*, **100**(26), pp. 15643–15648.

Question 6.2.1: B

This experiment didn't test what the periwinkles preferred to eat (A), so we can't tell if this is the case. This experiment also didn't assess the growth rate of periwinkles eating different food (C). Answer (D) is unlikely given that we are told that periwinkles defecated on the wounds they created but also grew larger on these leaves.

Question 6.2.2

The periwinkles were 'farming' the fungi. It could be that they actually 'seeded' the damaged leaves with fungal spores present in their faeces. Another example of fungal farming is from tropical ants which grow fungi in their nests on leaves collected by the ants.

Question 6.2.3

The fungi were able to grow on the damaged leaves without being eaten by periwinkles.

Question 6.2.4

The most likely reason is that there were more fungi initially on this material, so the snails were able to eat more and grow larger.

Exercise 6.3

Morton, J.P. and Silliman, B.R., 2020. Parasites enhance resistance to drought in a coastal ecosystem. *Ecology*, **101**(1), pp. e02897.

Question 6.3.1: C

Table 6.3 shows that the addition of salt (drought simulation) was a highly significant factor in both leaf damage and above-ground biomass; Figure 6.3 gives the detailed results. For leaf damage (radulation length), there was no difference between salt addition or not when periwinkles were infected with

Parorchis; there was only a significant difference when the periwinkles were uninfected. For above-ground biomass, there was a significant reduction here when salt was added, so answer A is incorrect. As set out already, there is a reduction in biomass due to salt addition, so B is also wrong. C is correct because there was certainly greater above-ground biomass with increasing parasite infection, for both with and without salt (although there was no difference between 0 and 10 parasites, the overall trend was for less saltmarsh cordgrass eaten as parasite load goes up). Similarly for radulation length, there was no difference between infection with 10 or 30 parasites in the drought treatment, but otherwise, the results with salt added and the no-addition treatment showed the same trend of reduced damage from grazing.

Question 6.3.2

The only pairwise comparison that showed a significant difference between salt addition and no added salt was when periwinkles were uninfected. The other pairs of bars for 10% and 30% infection rates showed no significant difference between no salt or salt addition. The significant interaction term showed that simulated drought only had an effect in the absence of parasites.

Question 6.3.3: D

A, B and C are all factually correct observations from the graph, but they don't explain why there was no significant interaction. Although parasite prevalence affected the above-ground biomass (hence both salt addition and parasite prevalence were significant as main effects), the pattern remained the same with significantly higher above-ground biomass when no salt was added compared with when there was salt added.

Question 6.3.4

Periwinkles which stay at the base of the plant are more likely to be underwater at high tide. Parasite larvae may spread through the water or periwinkles may get washed to new areas, spreading the parasite with them.

Exercise 6.4

Peng, S., Moret, A., Chang, C., Mayfield, A.B., Ren, Y., Chen, W.U., Giordano, M. and Chen, C., 2020. A shift away from mutualism under food-deprived conditions in an anemone-dinoflagellate association. *PeerJ*, **8**, p. e9745.

Question 6.4.1: D

This relationship is tricky to pin down. It wasn't an obligate mutualism (A) because the anemone did well without the symbiont when there was food available. Answer B could be correct under some circumstances, but as a bold statement without any additional detail, this doesn't actually sum up the relationship. C appears to be right, but, even under fed conditions, the anemone still grew larger when *Breviolum* was present so there must have been some benefit or the anemone wouldn't share its resources with the symbiont. That leaves D, a more nuanced statement—the relationship changed according to circumstances.

Question 6.4.2
Breviolum may have switched behaviour to act as a parasite and used the disintegrating host tissue for food. Alternatively, the relationship could be commensal, where *Breviolum* benefited but there was no negative effect (from *Breviolum*) on the host organism.

Question 6.4.3
There are a few options: doubling the size of the experiment and keeping a set of each treatment in the dark to prevent photosynthesis; labelling experiments using ^{13}C and/or ^{15}N might also help to trace movements of nutrients between the partners.

Exercise 6.5
Bell, S.S., McElmurray, P., Creed, R.P. and Brown, B.L., 2024. Impacts of invasion on a freshwater cleaning symbiosis. *Oecologia*, **205**(3), pp. 669–680.

Question 6.5.1: D
Answer A is incorrect because that isn't what the graph shows (it may or may not be true). B is wrong because the symbionts weren't measured on the invasive crayfish (this is stated in the figure legend). C is wrong because decreased proportion of invasive crayfish increased symbiont diversity and abundance. That leaves D, which gives the correct interpretation.

Question 6.5.2
The symbiont dispersed during direct contact between crayfish. The invasive crayfish was more active and so more likely to encounter other crayfish, allowing dispersal of symbionts. The invasive crayfish was more efficient at removing (and thus killing) the symbiont, so the direction of dispersal will be from native to invasive. The symbiont reduced crayfish mortality, so fewer symbionts will have a negative effect on native crayfish.

Question 6.5.3
Without intervention, the native crayfish is likely to go locally extinct. In the longer term, the branchiobdellidan worms may also go extinct as they cannot survive alone. In the absence of the cleaning services of the worms, the invasive crayfish may experience a slower growth rate and increased mortality. However, this is probably not sufficient to drive the invasive species to extinction.

Exercise 6.6
Pellmyr, O. and Huth, C.J., 1994. Evolutionary stability of mutualism between yuccas and yucca moths. *Nature*, **372**(6503), pp. 257–260.

Question 6.6.1: Mutualism
Both parties received a net gain from the interaction.

Question 6.6.2: Parasitic interaction
The moths gained (more young), but the plant lost out as more seeds were eaten.

Question 6.6.3: B

The plant clearly had some mechanism for detecting numbers of eggs laid (A), although it is not clear what this is. It wasn't physical damage to the outside of the ovary by oviposition and it wasn't damage inside the ovary by the feeding larva as the flower was aborted before this stage. The selection–abortion hypothesis states that when resources are insufficient for all seeds to mature, it is more likely that seeds with higher numbers of moth eggs will be aborted, rather than the reverse (C). Selection for retaining flowers with more moth eggs could lead to runaway increases in moths (D); this interaction is a fairly delicate balance of costs and benefits to both organisms. The most likely reason is insufficient pollination (B); remember this is mutualism so there had to be a benefit to the plant as well as the moth.

Question 6.6.4

Aborting flowers with more than 12 eggs was a 'check' on the mutualism which prevented it switching to parasitism. There was no value to the plant being pollinated by moths if all the seeds were subsequently eaten by moth larvae.

Exercise 6.7

Pennings, S.C. and Callaway, R.M., 1996. Impact of a parasitic plant on the structure and dynamics of salt marsh vegetation. *Ecology*, 77(5), pp. 1410–1419.

Question 6.7.1: D

The control plots showed the situation in the absence of dodder, and pickleweed always had the highest biomass. Even where dodder was present, in three out of the four combinations, pickleweed still had the greatest biomass. There is no data on the biomass of dodder itself here so we can't compare that against pickleweed (A). Plus, as dodder parasitises pickleweed, we wouldn't usually describe it as being competitively superior or inferior to its host plant.

Question 6.7.2

Large patches were described as having dead pickleweed in their centre; this presumably allows the other species to establish/increase. In small patches of dodder, there was still too much live pickleweed.

Question 6.7.3: A

Where there were large patches of dodder on the high marsh, there was more California sea lavender, so this would show a positive correlation; but this association was only apparent for large patches on the high marsh.

Question 6.7.4

Environmental conditions on the low marsh may naturally limit the growth of these species.

Question 6.7.5

When abundant (i.e. large patches), dodder had a strong negative impact on the competitively dominant pickleweed, reducing its biomass. This allowed less competitive species (California sea lavender and alkali heath) to establish and increase in size. Dodder is an annual and mostly dies after flowering. This left bare patches in which California sea lavender and alkali heath could persist; dodder will not grow in these areas as there was not enough of its favoured host species. Pickleweed will re-establish in these areas after a few years, reducing the other species. This allows dodder to colonise and the cycle will repeat.

Exercise 6.8

Cimadom, A. and Tebbich, S., 2021. Timing of infestation influences virulence and parasite success in a dynamic multi-host–parasite interaction between the invasive parasite, *Philornis downsi*, and Darwin's finches. *Oecologia*, **195**(1), pp. 249–259.

Question 6.8.1: B

Question 6.8.2: C

Question 6.8.3: A

Question 6.8.4: A, B, G

Ease of finding hosts (B), which was linked with how common a host is (A), could both affect host choice by female flies (although prevalence at the nestling stage was high and the same for both species). Host tolerance (C) is more complicated. For a parasite that uses more than one host, it can drive one species to extinction without having too much detrimental effect on itself, plus it hasn't had very long (~60–70 years) for much adaptation to its hosts (E). Small tree finch nests were more likely to fail, but this was still the current favoured species, which suggests that answer (D) wasn't a key factor. The data don't suggest that female flies took any account of number of flies already present (F). Blood quality may be important (G); small tree finches are omnivorous, and this could affect their blood chemistry in a way that is positive for vampire flies, but this wouldn't have changed over the last 30 years, so doesn't explain the switch by flies to parasitise this species more.

Question 6.8.5

The prevalence of vampire flies in green warbler finch nests between 1998 and 2005 was significantly higher than in 2012–2017. Small tree finch nests had significantly more larvae in 2012–2017, but nest size is unlikely to have changed over this time.

Question 6.8.6

Currently, during the incubation phase, more vampire fly eggs were laid in warbler finch nests and larvae presumably fed on the incubating adult bird. Flies might switch to laying more eggs in small tree finch nests to provide a longer time feeding on their preferred host.

Chapter 7

Exercise 7.1

Mikkelsen, A.J., Zedrosser, A., Sergiel, A., Hobson, K.A., Selva, N. and Hertel, A.G., 2025. Unraveling omnivory and community interactions between primary producers and an apex predator. *Ecology and Evolution*, **15**(4), p. e71181.

Question 7.1.1

It made no difference; bilberry production was highly variable and may show a slight decline while the proportion of bilberry in the bear diet may show a slight increase.

Question 7.1.2

It made no difference; there might have been a slight decrease in moose in the bear diet.

Question 7.1.3

Bilberries are available for longer than newly born calves, so are a more reliable food; bilberries don't run away or require long handling times as will be the case with moose calves.

Exercise 7.2

Werner, J.R., Gillis, E.A., Boonstra, R. and Krebs, C.J., 2016. You can hide but you can't run: Apparent competition, predator responses and the decline of Arctic ground squirrels in boreal forests of the southwest Yukon. *PeerJ*, **4**, p. e2303.

Question 7.2.1: C

If ground squirrels were competing with hares for either food (A) or territory (B), we would expect to see the population of one species increasing when the other decreased, rather than following the more synchronous pattern seen here. Predator satiation (D) probably drove the sharp increases in both prey species but not the decreases. C is the correct answer; reduced predation on both species allowed them to increase when predator numbers were low, but as predator populations increase, first hares, then ground squirrels were negatively affected, and the higher populations of predators will drive down prey populations very rapidly.

Question 7.2.2

The population appeared to have a threshold density of around 0.4 ha^{-1}, below which the population did not increase. This could be due to Allee effects such that reduced vigilance and predator detection resulted in a disproportionate impact by predators. It could also be due to changes in predator behaviour such as prey switching or additional food source availability, for example, over winter, allowing higher populations of predators to survive until the following spring.

Exercise 7.3

Watari, Y., Takatsuki, S. and Miyashita, T., 2008. Effects of exotic mongoose (*Herpestes javanicus*) on the native fauna of Amami-Oshima Island, southern Japan, estimated by distribution patterns along the historical gradient of mongoose invasion. *Biological Invasions*, **10**(1), pp. 7–17.

Question 7.3.1: A, B, C, D, E, F, H

For all these species, there was a positive relationship (shown with fitted lines) between presence or abundance and distance from the release point (i.e. density of mongooses).

Question 7.3.2: No

Question 7.3.3

The mongooses appeared to prey on larger animals which may eat crickets and cockroaches. Crickets and cockroaches would then experience reduced predation pressure and indirectly benefit from the mongoose.

Exercise 7.4

Barrios-O'Neill, D., Bertolini, C. and Collins, P.C., 2017. Trophic cascades and the transient keystone concept. *Biological Conservation*, **212**, pp. 191–195.

Question 7.4.1: C, E

There are no asterisks between the two dark bars and between the two white bars in the crab movement graph, showing there was no difference between activity due to reef presence/absence (A, B). There were big differences between the dark and white bars in the left-hand graph, confirmed by the asterisks showing significant differences between the presence and absence of skate (C). Answer D is correct in the absence of reefs, but when reefs were present it made no difference whether skate were present or absent. When there was a reef, the presence or absence of flapper skate made no difference to mussel recruitment (E). Only skate or a reef was needed, not both: in the absence of a reef, flapper skate made a big reduction to mussel mortality, but where there were reefs, there was no difference in mussel mortality when skate were present or absent (F).

Question 7.4.2

Flapper skate ate crabs; in addition, crab movement reduced when skate were present, presumably to avoid being eaten. Reduced crab activity reduced mussel mortality. Where a reef was present, there were more hiding places for mussels to attach, making it more difficult for crabs to predate on them.

Question 7.4.3

Conservation activity could either focus on increasing populations of large elasmobranchs, or artificial reef structures could be created.

Exercise 7.5

Willson, J.D., 2017. Indirect effects of invasive Burmese pythons on ecosystems in southern Florida. *Journal of Applied Ecology*, **54**(4), pp. 1251–1258.

Question 7.5.1: C

Answer A is incorrect because the relationship that Willson was investigating was between turtles and mammals (and only indirectly, pythons). It's unlikely that pythons would be able to dig up the turtle nests, plus the pythons had an indirect effect on turtles, so B is wrong. Pythons are unlikely to eat many, if any, turtles (D). Whilst E is correct (although it doesn't appear that pythons are suffering much from habitat destruction), this is rather vague and doesn't really answer the question.

Question 7.5.2: C

Answer A is correct but could be more precise (what sort of relationship is it?). From these data, we can't tell if B is correct; there were more predation events, but we don't know what was eating the eggs (at this stage). C is the most detailed, correct explanation; the relationship was positive and it was significant. D is correct but misses that the relationship was positive.

Question 7.5.3: C

Turtle nest predation wasn't highest in the core areas; it was actually lowest in these areas (A). There weren't (yet) any pythons in the extra-limital areas, so B can't be the correct answer. Turtles weren't directly benefiting from pythons; it was an indirect relationship (D), plus the way this is worded makes it sound as if the turtles and pythons had come to some sort of friendly arrangement.

Exercise 7.6

Yoshida, T., Jones, L.E., Ellner, S.P., Fussmann, G.F. and Hairston, N.G., 2003. Rapid evolution drives ecological dynamics in a predator–prey system. *Nature*, **424**(6946), pp. 303–306.

Question 7.6.1: C

This isn't really showing a 'typical' predator–prey cycle (A), as the time lag was such that when one species was at its peak, the other was at its trough. B seems unlikely because by the time rotifers were at their highest, algae had already crashed. D might happen, but we don't have any data to support this idea and there is nothing in the information to suggest that an alternative food was available to the rotifers.

Question 7.6.2: A, C

Answer A is likely to be the case and explains why this effect was only apparent in mixed-genotype cultures. If B is true, it's not clear how that would lead to the result we see. Predation will act as a strong selective pressure in favour of resistant genotypes, so C is correct and D is wrong.

Question 7.6.3

Rotifer predation selected for algae that had defences to grazing, but there was probably a cost to these defences. When rotifer populations were low, the cost of producing defences was greater than the advantage, so these genotypes were out-competed by less resistant genotypes. This allowed rotifers to increase again.

Exercise 7.7

Espelta, J.M., Arias-Leclaire, H., Fernández-Martínez, M., Doblas-Miranda, E., Muñoz, A. and Bonal, R., 2017. Beyond predator satiation: Masting but also the effects of rainfall stochasticity on weevils drive acorn predation. *Ecosphere*, **8**(6), p. e01836.

Question 7.7.1: A

Producing more seeds than the resident predators can eat is a good way to ensure that some survive (A). Climate certainly plays a role (B), so this is partly right, but there's more going on than this. Trees may need to recover between masting events, but this doesn't explain why they do this in the first place (C); why not just produce fewer seeds every year? D is true but doesn't actually explain why it's a good strategy.

Question 7.7.2: C

There appears to be a strong relationship between early autumn rain and percentage acorn predation (A), but this is an artefact of how the graph has been constructed; these two variables are on different scales (axes), and if these had been constructed differently, there would be less apparent correlation. There wasn't a clear pattern between acorn production and autumn rainfall (B). For example, in 2002 and 2008, rainfall was low but acorn production was very high. This makes sense as acorn development would start much earlier in the summer, so autumn rainfall is unlikely to affect this. The more acorns that were produced, the lower the proportion that was eaten (C). There were some patterns here, so answer D isn't right.

Question 7.7.3

Autumn rainfall in 2007 was very low; Figure 7.7b shows that there was a strong relationship between the numbers of weevils emerging and rainfall. This suggests that there were probably smaller numbers of weevils present in 2007 and thus the proportion of acorns eaten was lower.

Exercise 7.8

Morin, P.J., 1983. Predation, competition, and the composition of larval anuran guilds. *Ecological Monographs*, **53**(2), pp. 119–138.

Question 7.8.1: Eastern spadefoot toad

Question 7.8.2: Cope's gray treefrog

Question 7.8.3

Without predatory newts, eastern spadefoot toad was competitively dominant and Cope's gray treefrog was a subordinate member of the community. When eastern spadefoot toad numbers were reduced, the treefrog was able to out-compete the remaining amphibians. To help it survive newt predation, the treefrog may be camouflaged, or have avoidance strategies or chemical defences.

Chapter 8

Exercise 8.1

Schmitz, O.J., 2006. Predators have large effects on ecosystem properties by changing plant diversity, not plant biomass. *Ecology*, **87**(6), pp. 1432–1437.

Schmitz, O.J., 2003. Top predator control of plant biodiversity and productivity in an old-field ecosystem. *Ecology Letters*, **6**(2), pp. 156–163.

Question 8.1.1

See Figure 10.1.

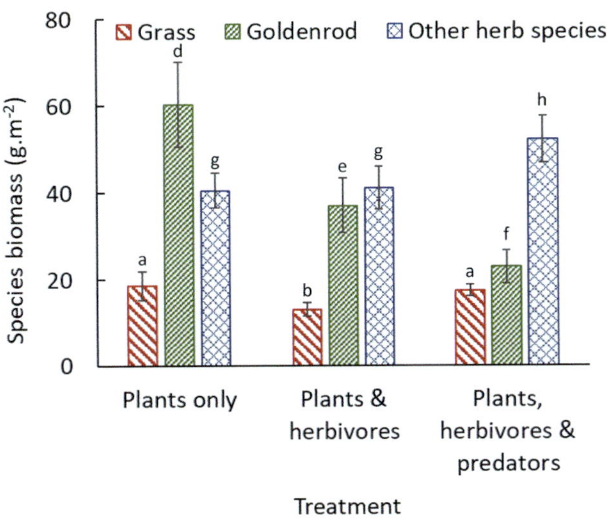

Figure 10.1 Biomass of each plant species when there is no herbivory, only herbivores, or herbivores and predators present. Bars show means ± 1 SD. (Adapted from Schmitz, 2006. Reproduced with permission, copyright 1983 by the Ecological Society of America.)

Question 8.1.2: B

Question 8.1.3: A, B

Question 8.1.4: C

Question 8.1.5: B
Table 8.1 shows that primary productivity was highest in the absence of grazing and lowest when the predator was present. By a process of elimination, goldenrod must have been responsible for most of the primary productivity as there was little change in the biomass of grasses; other herbs were increasing, but goldenrod decreased as grasshoppers and then spiders entered the system.

Question 8.1.6
Grasshoppers ate grasses and goldenrod, the competitively dominant species. When spiders were present, grasshoppers switched to feed mainly on goldenrod, probably because the denser growth provided better protection from spiders. When goldenrod decreased, other herb species were less suppressed and grew more.

Exercise 8.2

Ramirez, J.I., Poorter, L., Jansen, P.A., Den Ouden, J., Siewert, M. and Olofsson, J., 2023. Top-down and bottom-up forces explain patch utilization by two deer species and forest recruitment. *Oecologia*, **201**(1), pp. 229–240.

Question 8.2.1: A, B
Top-down factors are ones from higher trophic levels that affect the deer, so here, the impact of people acting as predators (Hunting) and through disturbance factors (Town).

Question 8.2.2: E, F, G
Bottom-up forces are those from lower trophic levels (resource availability) which affect patch use by deer.

Question 8.2.3: A, D, E
Patch use by roe deer had a significant (negative) relationship for bottom-up effects (patch quality) and a significant (negative) relationship with hunting. Patch use by red deer only had a significant (negative) relationship with hunting.

Question 8.2.4
Red deer are a social species that live in herds, whereas roe deer are solitary.

Question 8.2.5: C
Red deer do live in herds and probably do experience competition for food (A), but the data analysis showed that patch quality was more important for roe than red deer. Roe deer do have a more restricted diet than red deer, and this might affect the areas in which they choose to feed (B), but it doesn't explain why they would spend less time in better-quality areas. In high-quality areas

they can find food quickly (C) and then move on to other areas. The analysis suggests that roe deer would spend more time in poor-quality areas (i.e. higher proportion of conifers), but we don't know whether that would be to avoid predators or just because it takes longer to find enough food.

Question 8.2.6

Greater numbers of deer led to higher patch utilisation; higher patch utilisation led to a greater, negative impact on patch quality (for roe deer only).

Exercise 8.3

Spiller, D.A. and Schoener, T.W., 1990. A terrestrial field experiment showing the impact of eliminating top predators on foliage damage. *Nature*, **347**(6292), pp. 469–472.

Question 8.3.1: A

Figure 8.3a shows that there were far more leaf scars than galls, so this eliminates B, which shows all interactions being of equal strength. Removing lizards led to an increase in scars (so more homopterans) and a decrease in galls (so fewer midges). But lizards also ate spiders, so these increased in the absence of lizards. The decrease in midges and increase in homopterans when spiders increase suggests that spiders had more impact on gall midges than they did on homopterans, so we can reject C, which shows the strength of these two interactions the other way around, and D, which shows the strength of these interactions as the same. This is a tricky question as we don't know whether lizards had the same impact on homopterans as they did on spiders, i.e. in the absence of lizards, did the increase in spiders affect homopterans to the same extent? If this were the case, we would expect to see no difference in Figure 8.3a for homopterans when lizards were present or absent.

Exercise 8.4

Everatt, K.T., Andresen, L., Ripple, W.J. and Kerley, G.I.H., 2016. Rhino poaching may cause atypical trophic cascades. *Frontiers in Ecology and the Environment*, **14**(2), pp. 65–67.

Question 8.4.1

See Figure 10.2.

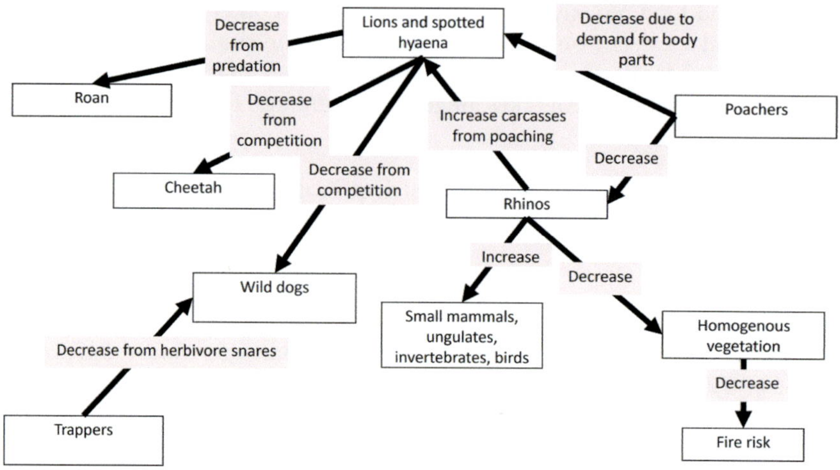

Figure 10.2 Completed interaction diagram from Kruger National Park. (Adapted from Everatt et al., 2016. Reproduced with permission, copyright 2016 The Ecological Society of America.)

Exercise 8.5

Wootton, J.T., 1992. Indirect effects, prey susceptibility, and habitat selection: Impacts of birds on limpets and algae. *Ecology*, **73**(3), pp. 981–991.

Wootton, J.T., 1993. Size-dependent competition: Effects on the dynamics vs. the end point of mussel bed succession. *Ecology*, **74**(1), pp. 195–206.

Question 8.5.1: C

For the communities where birds had access to the rocks, the total cover by both species combined was below 50% at the start of the succession, but within one year goose barnacles started to decline. It's unlikely that competition from mussels could have this effect so early (A, B). Where birds were excluded, the goose barnacles showed an initial strong increase (C), whereas mussels increased in both the treatments with and without birds (D). The areas chosen for this experiment were dominated by goose barnacles, so the greater cover by this species at the start of the experiment does not demonstrate differences in colonising ability (E).

Question 8.5.2: A

Birds did not seem to eat shield limpet, so B is incorrect. Checkered limpet increased when birds were present, so C is also wrong. Given that the cover of shield limpet didn't change, it seems unlikely that the decrease in ribbed limpet was caused by competition between these species. Ribbed limpet could have been responding to the decrease in cover by checkered limpet (E), but the increase in the former was considerably greater than the decrease in the latter, so on balance, the most likely answer is A.

Question 8.5.3: D

Given that shield limpets increased in the absence of birds but that ribbed limpet increased substantially when birds are excluded, the most likely reason for this was competitive exclusion by ribbed limpet (D).

Question 8.5.4

Each limpet species had a selective advantage to settling on its preferred substrate species.

Question 8.5.5: A, E

Ribbed limpet is pale and camouflaged on pale goose barnacles; shield limpet is dark and camouflaged on mussels.

Question 8.5.6: C

Gulls and oystercatchers don't eat algae; although not explicitly stated elsewhere in the question, this is implied by the earlier results (A, B, D), and algal cover increased, so B is wrong in any case. If birds competed with limpets, Figure 8.5b would look different—in the absence of birds, all the bars would increase. E could be correct, as it follows the same logic as C, but algae was at such a lower percentage cover, even in the absence of birds, that it seems unlikely that it competed for space with the barnacles and mussels.

Question 8.5.7: A

It is likely to be ribbed limpet as this was the species with the strongest response when birds were excluded.

Question 8.5.8

See Figure 10.3.

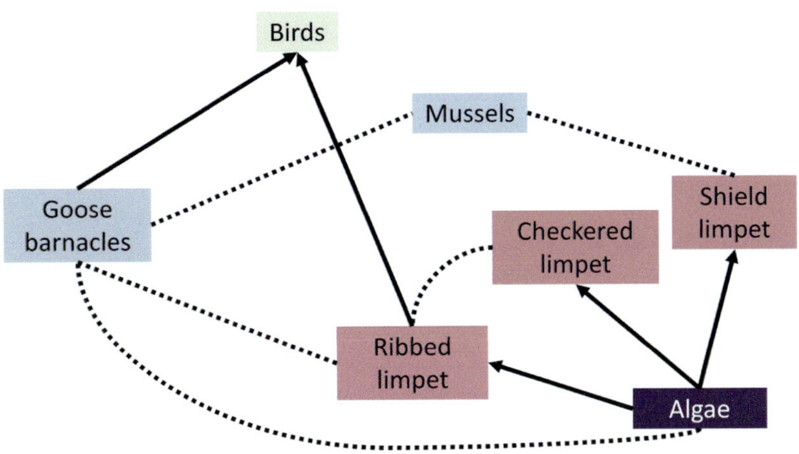

Figure 10.3 Connections between the main species in this coastal community. Solid arrows indicate direct interactions with the arrow showing direction of energy transfer, for example species D eats algae. Dotted lines show indirect interactions between species.

Question 8.5.9
See Figure 10.4.

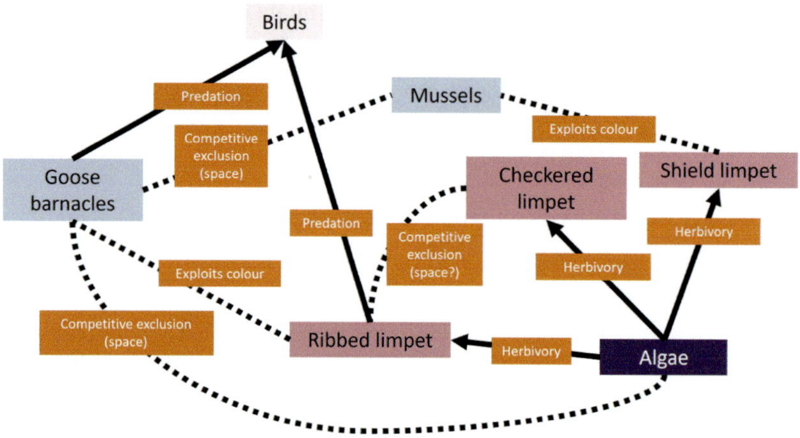

Figure 10.4 Connections between the main species in this coastal community and interactions. Solid arrows indicate direct interactions with the arrow showing direction of energy transfer, for example species D eats algae. Dotted lines show indirect interactions between species.

Exercise 8.6
Wenting, E., Jansen, P.A., Burggraeve, S., Delsman, D.F., Siepel, H. and Van Langevelde, F., 2024. The influence of vertebrate scavengers on leakage of nutrients from carcasses. *Oecologia*, **206**(1), pp. 21–35.

Question 8.6.1: A
Option B is less likely to be the case as vertebrates will move nutrients away from the site of the carcass via their excrement and the energy incorporated into their bodies, so a lack of vertebrates will tend to keep nutrients in the same location. However, the lack of opening up the carcass is only going to be a short-term issue. Areas around carcasses don't usually end up without any vegetation; the reverse is more likely.

Question 8.6.2: A, C
The green line for 'all mammals and birds excluded' showed greater Zn than the other treatments (A). All other treatments showed no difference (B, D).

Question 8.6.3
Wild boar compete with other scavengers. In the absence of wild boar, other species were able to use more of the carcass.

Question 8.6.4
The concentration of elements might have remained the same, but the biomass of the plant could have increased in response to additional trace elements

available in the soil. However, if these plants were grazed, they might have been the same size, despite having taken up some of the elements.

Question 8.6.5

Flies may be important in some habitats and also result in nutrients being distributed more widely across an area.

Exercise 8.7

Fablet, L., Pellerin, A., Zarzoso-Lacoste, D., Dubut, V. and Walch, L., 2024. Metabarcoding reveals waterbird diet in a French Ramsar wetland: Implications for ecosystem management. *Knowledge & Management of Aquatic Ecosystems*, (425), p. 9.

Question 8.7.1: C

Coot had the longest bar on the right-hand side, showing that more samples came from this species.

Question 8.7.2: C

It's coot again and it had the highest number of links to the food items on the left-hand side.

Question 8.7.3: D

Non-biting midges had the greatest number of connections to species on the right-hand side of the diagram.

Question 8.7.4

The animals eaten by the herons may have fed on these plants, or faecal samples could have been contaminated with pollen or small pieces of plant litter.

Question 8.7.5: B, E

Blackbirds and mallard both ate 12 species, closely followed by coots and wood pigeons with 11.

Question 8.7.6: A, C, G

Mute swans had three plant items in their diet; moorhens and grey herons both had five.

Question 8.7.7: Plants

Linkage is calculated by Number of links/Number of species. For plants: 66/39 = 1.69. For invertebrates: 33/24 = 1.38.

Exercise 8.8

Calderón-sanou, I., Münkemüller, T., Zinger, L., Schimann, H., Yoccoz, N.G., Gielly, L., Foulquier, A., Hedde, M., Ohlmann, M., Roy, M., Si-moussi, S. and Thuiller, W., 2021. Cascading effects of moth outbreaks on subarctic soil food webs. *Scientific Reports*, **11**(1), p. 15054.

Question 8.8.1

The black bars do not overlap the 0.0 line, showing that there is a 90% probability that there was a difference in diversity for that group between forests affected or not by moth outbreaks. Grey bars all overlap with the 0.0 line, suggesting there was no difference. [The data were analysed with Bayesian statistics, which uses probabilities with credible intervals rather than confidence intervals.]

Question 8.8.2: D

Ectomycorrhizal fungal diversity decreased, and there was an even split (18 each) of groups showing no change or increase in diversity, so you could argue that A and C are also correct as well as D.

Question 8.8.3

This reflects the change from woodland where most trees form relationships with ectomycorrhizal fungi to communities dominated by grasses which form relationships with arbuscular mycorrhizas.

Question 8.8.4

As the birch trees died following defoliation, the dead wood provided new habitat for saprotrophic fungi that would persist for considerable periods of time following the moth outbreak.

Question 8.8.5

The change from a woodland habitat with a shrub understory to one dominated by grasses will have reduced the amount of plant and associated animal material entering the system.

Question 8.8.6: B

No, the effect of defoliation was equally important at all trophic levels.

Chapter 9

Exercise 9.1

Couet, J., Marjakangas, E., Santangeli, A., Kålås, J.A., Lindström, Å and Lehikoinen, A., 2022. Short-lived species move uphill faster under climate change. *Oecologia*, **198**(4), pp. 877–888.

Question 9.1.1

Can species disperse in sufficient numbers? Do they have a sufficiently high reproductive rate? How specialised are they (generalists are likely to do better)? Are there suitable vacant niches, so that interspecific competition is minimised? Do species show high site fidelity? This isn't an exhaustive list!

Question 9.1.2: D

The points on the graph represent grid cells with the average change in altitude by birds, in metres. For answers A or B to be correct, the y-axis would need to represent numbers of birds. Answer C may or may not be the case, but it isn't what was measured here. Answer D is correct as both factors are measured in metres; it simply shows that if there was more altitudinal height available to move into, so birds moved into it.

Question 9.1.3: A

Option A is correct; it's a simple relationship between longevity and change in altitudinal range. For B to be correct, the line would need to be sloping upwards, i.e. a positive relationship. For C to be correct, the fitted line would need to be flatter and the confidence intervals overlapping with the 0 line. Answer D may be part of a correct reason for why this pattern occurs, but it's not what the graph shows.

Question 9.1.4

Longevity is strongly related to reproductive rate, so shorter-lived species may react more quickly to changes in climate and be more easily measured over the timescale studied here.

Exercise 9.2

Selbach, C. and Poulin, R., 2020. Some like it hotter: Trematode transmission under changing temperature conditions. *Oecologia*, **194**(4), pp. 745–755.

Question 9.2.1: B

Answer A could be true but it implies more conscious volition than cercariae are likely to have. Answer B seems the most likely explanation because it links with the distance travelled. Answer C is incorrect in this context as the experiment was only run with the intermediate host and the cercariae so there was no opportunity to assess competition for hosts. Answer D could be correct but we are not given any details on cercariae size, so it's less likely to be the most probable answer.

Question 9.2.2: No

The reduced survival time was offset by the increased output of cercariae.

Question 9.2.3

Maritrema poulini parasitises a very common organism (crustaceans) and can increase cercariae output and the distance they travel, suggesting that this species may have a considerable impact on their definitive host. This is less likely to be the case with *C. parvum*, as active dispersal is much less. There may be some increase in the impact that this species has on crustacea. Changes in *Apatemon* sp. and *Aporocotylid* sp. are much less likely to have an impact, as their definitive hosts (fish) are less common in the environment and are thus more difficult for the short-lived cercariae to find.

Question 9.2.4

Longer-term experiments in field conditions; impact of warming on host species' survival and behaviour; changes in parasite ability to detect hosts; and hosts' ability to avoid infections.

Exercise 9.3

Gehring, C.A., Mueller, R.C., Haskins, K.E., Rubow, T.K. and Whitham, T.G., 2014. Convergence in mycorrhizal fungal communities due to drought, plant competition, parasitism, and susceptibility to herbivory: Consequences for fungi and host plants. *Frontiers in Microbiology*, **5**, pp. 306.

Question 9.3.1: D

Figure 9.3a doesn't give us any information about the impact of herbivory. It shows changes in mycorrhizal diversity associated with trees experiencing different levels of herbivory, and this is what the question is asking. We can't make straightforward interpretations using 'increased' or 'decreased' effect, as the analysis doesn't tell us what the axes represent—we have to work that out using ecological knowledge of the experiment or community. Thus, answers A and B are not correct. The points representing droughted communities are almost on top of each other, showing these were now very similar to each other (answer D). We can infer statistical significance from the error bars, but answer C is wrong because the communities were more, rather than less, similar following drought.

Question 9.3.2

The responses of mycorrhizal communities to the effects of parasitism, herbivory or competition on host trees are spread along Axis 1 and appear in the same sequence for both high and low biotic interactions. This suggests that there was a differential response to each stressor, regardless of its intensity. [Imagine how these points would appear if there were no Axis 2 and all points appeared on a single line representing Axis 1.] However, there was a greater difference between mycorrhizal communities when stressors were low intensity, with reduced mycorrhizal diversity where stressors were high intensity.

Question 9.3.3: A

The points represent the diversity of mycorrhizal communities, so answer C can't be correct. We don't have any data here about the degree of mycorrhizal colonisation, so answer B can't be right. Again, the points represent mycorrhizal communities, so we don't have any data on interactions between stressors, so answer D can't be right. That leaves answer A—there wasn't any significant difference between the stressors on Axis 2, but there was a big difference relating to whether the plants and thus mycorrhizas were experiencing low or high intensity of the stressors.

Question 9.3.4

These stressors probably increase the plant's need for resources from the soil, via mycorrhizas, but at the same time, the stressors reduce the plant's ability to photosynthesise and provide resources to the mycorrhizal fungi.

Question 9.3.5

Under future climate change with more drought conditions, mycorrhizal communities may become less diverse and dominated by a few species. Currently we don't have enough data to know if this is a problem. A less diverse mycorrhizal community may not be a problem for trees if those mycorrhizas are better adapted to drought conditions.

Exercise 9.4

Mccormick, M.I., Watson, S. and Munday, P.L., 2013. Ocean acidification reverses competition for space as habitats degrade. *Scientific Reports*, **3**(1), p. 3280.

Question 9.4.1: B

Ambon damselfish were more aggressive, and this is likely to make them the competitively dominant species.

Question 9.4.2: A

There is a reversal in which species were more aggressive, suggesting a changeover in the dominance of the species.

Question 9.4.3: C

For both species at ambient and elevated CO_2, there was an increased effect from unbleached to bleached corals, but this was not statistically significant for any of the combinations, as shown by the letters above the bars, so C is the correct answer. The species responded in the same way to bleaching, so there wasn't a statistical interaction here.

Question 9.4.5

Lemon damselfish showed no change in distance ventured from the reef under elevated CO_2, but at ambient CO_2 they moved farther when corals were bleached. The opposite pattern was shown by Ambon damselfish: there was no difference between bleached and unbleached coral at ambient CO_2, but there was a significant increase in movement on bleached coral under elevated CO_2.

Question 9.4.6

Elevated CO_2 affected these species differently: the Ambon damselfish became less aggressive under elevated CO_2 and is likely to lose territory. This may reduce their survival and numbers across the reef. This species also moved farther away from the reef and may be more vulnerable to predation. However, other species, including predators, may also be differentially affected by elevated CO_2, so it isn't possible to predict outcomes at the community scale.

Exercise 9.5

Hanssen, S.A., Moe, B., Bårdsen, B., Hanssen, F. and Gabrielsen, G.W., 2013. A natural antipredation experiment: Predator control and reduced sea ice increases colony size in a long-lived duck. *Ecology and Evolution*, **3**(10), pp. 3554–3564.

Question 9.5.1: A

The data for eider populations where predators were removed showed a clear increase compared to the area where predators were not removed.

Question 9.5.2: Positive

The t-value for predator removal was positive (4.630), so as there was increased predator control, there were higher numbers of eider. To check if you had this correct, you could also refer to the graph.

Question 9.5.3: B, D

The t-value for ice concentration was negative (–2.573). This shows that increasing ice concentration had a negative effect on the response variable (eider population size), so eiders are likely to increase as the climate warms.

Question 9.5.4

As eiders dive to find mussels, increased sea ice probably makes this more difficult. Having accessible water for longer periods makes it easier for them to forage.

Question 9.5.5

There was no significant relationship between lagged population density and current population density in the eider population in the nature reserve; the graph shows a horizontal fitted line. One explanation could be that the presence of predators prevented the population from increasing. It had hit its carrying capacity; even if there were many more individuals the previous year, the population remained stable. However, there was a significant, positive relationship between these factors in the area where predators were removed, showing that if there were more individuals the previous year, the population increased in the following year.

Exercise 9.6

Mcfarlane, K.J., Cusack, D.F., Dietterich, L.H., Hedgpeth, A.L., Finstad, K.M. and Nottingham, A.T., 2024. Experimental warming and drying increase older carbon contributions to soil respiration in lowland tropical forests. *Nature Communications*, **15**(1), p. 7084.

Question 9.6.1

Increased CO_2 emissions from burning fossil fuels have increased the amount of $^{12}CO_2$ and thus further reduced $\Delta^{14}C$ (i.e. 'diluted' it).

Question 9.6.2: A, B, G

Check that you spotted which pairs of bars have an asterisk (*) showing they are significantly different.

Question 9.6.3

The CO_2 being respired was derived from older carbon, i.e. it was fixed from the atmosphere longer ago than the CO_2 respired in the control and in the dry season.

Question 9.6.4

The ^{14}C was being respired by the soil microbes using soil organic matter derived from CO_2 fixed several decades ago.

Question 9.6.5

As the climate continues to warm, 'older' carbon substrates in soil may be broken down by microbes, reducing stores of carbon in soil and increasing CO_2 in the atmosphere.

Exercise 9.7

Yurkowski, D.J., Hussey, N.E., Ferguson, S.H. and Fisk, A.T., 2018. A temporal shift in trophic diversity among a predator assemblage in a warming Arctic. *Royal Society Open Science*, **5**, p. 180259.

Question 9.7.1: B

All top predators increased their intake of forage fish: Arctic char from 29% to 66%, ringed seal from 12% to 49%, beluga from 15% to 35%, and Greenland halibut from 79% to 95%.

Question 9.7.2

They were flexible in their feeding preferences and responsive to increases in suitable prey.

Question 9.7.3

Arctic cod have decreased and capelin have increased, but capelin have increased by more than cod have decreased, so there are more forage fish available to predators. Capelin may also be easier to catch than other prey.

Question 9.7.4: D

The smaller area shown by joining the centroids of each ellipse (the black quadrilateral shape on each graph) shows that the diets of these predators had become more similar over time. This was mostly driven by changes in the diet of belugas and seals.

Question 9.7.5

There may be more competition between the species as they focus on the same prey species.

Exercise 9.8

Kordas, R.L., Donohue, I. and Harley, C.D.G., 2017. Herbivory enables marine communities to resist warming. *Science Advances*, **3**(10), p. e1701349.

Question 9.8.1: C

The effect size peaked in summer (June–July–August), with greater impact for the warming treatments. The warming treatment communities took longer to recover, but by the following spring (February–March–April), there was very little difference between the treatments.

Question 9.8.2

Without limpets, there was a greater difference in species present between the warmed and ambient treatments.

Question 9.8.3: B, C

The points for each treatment combination which are closer together show that there was less diversity in their successional trajectory, i.e. they were more similar to each other at the end of the experiment. Points which are very spread out show that within a treatment the end points were less similar, i.e. more diverse. 'Ambient, with limpets' and 'Warmed, with limpets' have a more diverse set of outcomes, whereas the treatments without limpets were much more similar.

Question 9.8.4

Through their grazing activities, the limpets created gaps on the warmed and ambient plates. Limpet numbers varied over time, so the timing of gap appearance will also have varied. This enabled different species to colonise gaps, i.e. algae could establish if gaps appeared in January, whereas barnacles could colonise if gaps appeared in April. Barnacles and large algae provided structure for other species to colonise, increasing diversity and thus reducing similarity to warmed plates where limpets were also present.

Question 9.8.5

Rocky shores with intact communities, including herbivores, may be better able to cope with the effects of climate change.

Index

A

Acacia drepanolobium, *see* whistling thorn
Acacia suaveolens, *see* sweet wattle, 21–23, 155–156
Acanthaster planci, *see* crown-of-thorns starfish
Acorn weevil (*Curculio* spp.) 105–106, 182
Acropora spp., 23–24, 156
African jewel fish (*Rubricatochromis letourneuxi*), 15–19, 153
Alces alces, *see* moose
Aldabra Atoll, Seychelles, 48–50, 163–164
Algae
  brown, 29–30, 158
  rotifers, 103–104, 181–182
Alkali heath (*Frankenia salina*), 90–91, 177–178
Alkali weed (*Cressa truxillensis*), 26–28, 157
Alkaloids, 55, 165–166
Allee effect, 179
Alternative stable state, 28–30, 158
Altitudinal range shift, 130–132, 190–191
Amami-Osima Island, Japan, 98–99, 180
Ambon damselfish (*Pomacentrus amboinensis*), 136–138, 193
Amensalism, 78–80, 173–174
American black oystercatcher (*Haematopus bachmani*), 116–121, 186–188
American mink (*Neogale vison*), 75–77, 172–173
Anas spp., *see* duck
ANOVA, repeated measures, 69–70, 170
Ant
  Crematogaster mimosae, 78–80, 173–174
  red wood (*Formica polyctena*), 71–72, 170–171
  Scottish wood (*Formica aquilonia*), 71–72, 170–171
Apatemon sp., *see* trematode flatworm
Aporocotylid sp., *see* trematode flatworm
Aposymbiotic, 85–86, 175–176
Arctic char, 143–146, 195
Arctic cod, 143–146, 195
Arctic ground squirrel (*Urocitellus parryii*), 97–98, 179
Ascophyllum nodosum, *see* knotted wrack
Atlantic puffin, 72–73, 171
Average xvi
Avian vampire fly (*Philornis downsii*), 91–94, 178

B

Barnacle
  barnacle sp., 146–148, 196
  common (*Semibalanus balanoides*), 28–30, 158
  goose (*Pollicipes* spp.) 116–121, 186–188
Beetle community, 8–9, 151–152
Beluga whale, 143–146, 195
Bilberry (*Vaccinium myrtillus*), 95–97, 179
Biodiversity
  bird communities, 11–13, 152–153
  colonisation of islands, 8–9, 151–152
  compensatory effects, 1–4, 150
  complementarity, 1–4, 150
  ecosystem function, 1–4, 150
  fynbos, 19–20, 155
  grasslands, 14, 153

herbivory, 9–11, 11–13, 152, 152–153
invasibility, 6–8, 151
resilience, 1–4, 150
stability, 1–4, 150
Biplot xx–xxiii
Blackbird, 127–127, 189
Black cherry (*Prunus serotina*), 9–11, 152
Blue mussel (*Mytilus trossulus*), 116–121, 186–188
Bottom-up processes, 72–73, 113, 170, 184
Box spp. (*Buxus* spp.) 53–56, 165–166
Brachypodium pinnatum, *see* heath false brome
Branchiobdellida, *see* crayfish worm
Bray-Curtis similarity index, 146–148, 196
Breviolum minutum, *see* dinoflagellate
Bromus tectorum, *see* cheatgrass
Brown bear (*Ursus arctos*), 95–97, 179
Burmese python (*Python bivittatus*), 101–103, 181
Burning management, 19–20, 21–23, 155, 155–156
Buxus spp, *see* box spp.

C

^{13}C
Arctic food webs, 143–146, 195
brown bear diet, 95–97, 179
sea anemones, 85–87, 176
seabirds, 72–73, 171
^{14}C, 141–143, 194–195
C:N ratio, 49, 164
Caddis fly (*Pycnopsyche* spp), 36, 159–160
California sea lavender (*Limonium californicum*), 90–91, 177–178
Calluna vulgaris, *see* heather
Camarhynchus parvulus, *see* finch, small green
Cambarus spp, *see* crayfish spp.
Canary Islands, 8–9, 151–156
Capelin, 143–146, 195
Capitella capitata, 51–53
Capreolus capreolus, *see* deer, roe
Carcass, 122–124, 188–189
Carnivory, 8–9, 58–61, 151–152, 166–167
Carrying capacity (K), 52–53, 164–165
Centaurea maculata, *see* spotted knapweed
Centaurea stoebe, *see* spotted knapweed
Ceratadon purpureus xxi
Cercariae, 132–134, 191–192
Certhidea olivacea, *see* finch, green warbler
Cervus elaphus, *see* deer, red
Character displacement, 72, 73, 171
Cheatgrass (*Bromus tectorum*), 31–32, 158–159
Cheetah, 115–116, 185–186
Chi square test, 71–72, 170–171
Climate change
Arctic food webs, 143–146, 195
birds' altitudinal range, 130–132, 190–191
coastal communities, 146–148, 196
damselfish, 136–138, 193
eider 139–141, 194
mycorrhizal community, 134–135, 192–193
parasitic trematodes, 132–134, 191–192
pink muhly grass, 66–67, 169
tropical forest soils, 141–143, 194–195
CO_2
ambient, 136–138, 193
elevated, 136–138, 193, 194–195
Coccoloba uvifera, *see* sea grape
Coefficient of variation (CV), 3, 150
Coexistence
bluebirds, 67–69, 169
ducks, 73–74, 171–172

fugitive, 71–72, 171
wood ants, 71–72, 170–171
Coitocaecum parvum, *see* trematode flatworm
Collserola, Spain, 105–106, 182
Columba palumbus, *see* wood pigeon
Commensalism, 78–80, 162, 176
Common cottongrass (*Eriophorum angustifolium*) xvi
Common murre, 72–73, 171–172
Compensatory effects, 1–4, 150
Competition-colonisation, *see* trade off
Competitive dominance, 23–24, 156, 192–193
Competitive exclusion
grasslands, 14, 153
limpets, 116–121, 186–188
seabirds, 72–73, 171
Competitive release, 72, 73, 171
Complementarity, 1–4, 150
Confidence interval xv–xx
Consumer, 34–37, 158–159
Cooperation, 73
Coot, 124–127, 189
Coral
bleaching, 136–138, 193
community xxiii
reef disturbance, 23–24, 25–26, 156, 156–157
spawning, 39–40, 161
Correlation coefficients, 2
Crayfish spp. (*Cambarus* spp.) 87–88, 176
Crayfish worm (Branchiobdellida), 87–88, 176
Crematogaster mimosae, *see* ant
Cressa truxillensis, *see* alkali weed
Crown-of-thorns starfish (*Acanthaster planci*), 23–26, 156–157
Cumberland Sound, Nunavut, Canada, 143–146, 195
Curculio spp., *see* acorn weevil
Cuscuta salina, *see* marsh dodder
CV, *see* coefficient of variation
Cyclone Oli, 23–26, 156–157
Cydalima perspectalis, *see* moth, box tree

D

Daphnia longispina, 69–70,170
Decomposition
carcasses, 122–124, 188–189
food webs, 122–124, 188–189
home-field advantage, 42–43, 161
litter bags xvi–xx, 43–45
plant litter, 42–43, 162
soil communities, 127–129, 189–190
Deer
red (*Cervus elaphus*), 111–114, 184–185
roe (*Capreolus capreolus*), 111–114, 122–124, 184–185, 188–189
Denseflower cord grass (*Sporobolus montevidensis*), 27–28, 157
Dinoflagellate (*Breviolum minutum*), 85–87, 175–176
Dipturus intermedia, *see* flapper skate
Dispersal, 72, 171, 190–191
Distichlis spicata, *see* seashore salt grass
Disturbance
African jewelfish, 15–19, 153–154
coral communities, 23–24, 25–26, 156, 156–157
fynbos, 19–20, 155
Diversity
bird communities, 11–13, 152–153
mycorrhizal communities, 134–135, 192–193
D. magna, 69–70, 170
Drying effects on tropical soil, 141–143, 194–195
Duck (*Anas* spp.) 73–74, 171–172

E

Eastern newt (*Notophthalmus viridescens*), 106–107, 182–183
Ecotron, 4, 150
Eider (*Somateria mollissima*), 139–141, 194
Epirrita autumnata, *see* moth, autumn
Eriophorum angustifolium, *see* common cottongrass
Error bars, how to interpret xv–xx
Eurasian otter (*Lutra lutra*), 75–77, 172–173
European red ant (*Formica polyctena*), 71–72, 170–171
Everglades National Park, Florida, US, 15–19, 101–103, 180
Exaiptasia pallida, *see* sea anemone

F

Fagus japonica, *see* Japanese beech
Farming
  marsh periwinkles, 81–82, 174
  termites, 43–45, 162
Faxonius cristavarius, *see* spiny stream crayfish
Finch
  green warbler (*Certhidea olivacea*), 91–94, 178
  small tree (*Camarhynchus parvulus*), 91–94, 178
Fire
  fynbos, 19–20, 155
  seed germination, 21–23, 155–156
Fire-adapted community, 19–20, 21–23, 155, 155–156
Flapper skate (*Dipturus intermedia*), 99–101, 180
Food web
  Arctic, 143–146, 195
  bipartite, 124–127, 189
  linkage, 124–127, 189
  soil, 127–129, 189–190
Forage fish, 143–146, 195
Formica aquilonia, *see* ant, Scottish wood
F. polyctena, *see* ant, European red
Frankenia salina, *see* alkali heath
Freshwater turtle, 101–103, 181
Frigatebird, 48–50, 163–164
Fugitive coexistence, *see* coexistence
Fungi
  arbuscular mycorrhiza, 127–129, 189–190
  ectomycorrhizal, 127–129, 189–190
  saprotrophic, 127–129, 189–190
Fynbos plant community, South Africa, 19–20, 155

G

Galapagos finch, 91–94, 178
Galapagos Islands, 91–94, 178
Gall midge, 114–115, 185
Gause, 74–75, 172
Generalised least squared model, 53, 165–166
Generalist, 130–132, 190–191
General Linear Model (GLM), xvi–xix
Glaucous-winged gull (*Larus glaucescens*), 116–121, 186–188
Goldenrod (*Solidago rugosa*), 108–110, 183–184
Goniastrea spp., 23–24, 156
Grazing, *see* herbivory
Greenland halibut, 143–146, 195
Guano, 48–50, 163–164

H

Haematopus bachmani, *see* American black oystercatcher
Hawaii, US, 37–39, 160–161
Heath false brome (*Brachypodium pinnatum*), 14, 153
Heather (*Calluna vulgaris*) xvi
Herbivory

beetles, 8–9, 151–152, 166
bird communities, 11–13, 152–153
corals, 23–24, 25–26, 156, 156–157
elephants, 78–80, 173–174
grasshoppers, 108–110, 183–184
grasslands, 14, 153
homopterans, 114–115, 185
limpets, 116–121, 186–188, 196
mycorrhizal communities, 134–135, 192–193
rhino, 115–116, 185–186
roan, 115–116, 185–186
scale insects, 78–80, 173–174
termites, 43–45, 162
Heron Island, Great Barrier Reef, 39–40, 161
Hluhluwe-iMfolozi Park, South Africa, 43–45, 162
Holoparasite, 90
Home-field advantage, *see* decomposition
Homopteran, 114–115, 185
Horse mussel (*Modiolus modiolus*), 99–101, 180
H. ostralegus, *see* oystercatcher
Huecas, Spain, 105–106, 182
Hysteresis, 154

I

Indirect gain, 78–80, 173–174, 180, 181
Interaction, statistical
box tree moth, 55–56, 165–166
Daphnia, 69–70, 170
interpretation xvii, xix, xx
marsh periwinkle, 82–85, 174–175
Intermediate Disturbance Hypothesis, 19–20, 155
Intermediate host, 132–134, 191–192
Interspecific competition
American mink, 75–77, 172–173
apparent competition, 73
bluebirds, 67–69, 169
character displacement, 171
competitive dominance, 106–107, 182–183
competitive exclusion, 71–72, 170–171
competitive release, 171
Daphnia longispina, 69–70, 170
D. magna, 69–70, 170
Eurasian otter, 75–77, 172–173
European red ant, 71–72, 170–171
exploitation competition, 73
Gause, 74–75, 172
interference competition, 73
limpet species, 116–121, 186–188
mycorrhizal communities, 134–135, 192–193
niche partitioning, 171. 172
noisy miner 64–65, 168–169
Paramecium spp., 74–75, 172
pink muhly grass, 66–67, 169
resource partitioning, *see* niche partitioning
Scottish wood ant 71–72, 170–171
seabirds, 72–73, 171
temporal niche partitioning, 75–77, 173
Intraspecific competition
bird clutches, 61–62, 167
box tree moths, 53–56, 165–166
intergenerational, 53–56, 166
noisy miner 64–65, 168–169
oystercatcher 63–64, 168
seabirds, 72–73, 171
wood pigeon, 62–63, 167–168
Invasive species
African jewelfish 15–19, 153
avian vampire fly, 91–94, 178
Burmese python 101–103, 18
cheatgrass 31–32, 158–159
grassland succession, 31–32, 158–159
small Indian mongoose 98–99, 180
spiny stream crayfish, 87–88, 176
Isotopic niche, 143–146, 195

J

Janzen-Connell hypothesis, 9–11, 152
Japanese beech (*Fagus japonica*), 42–43, 161
Japanese hemlock (*Tsuga diversifolia*), 42–43, 161

K

K, carrying capacity, 51–53, 164–165
Kenya Long-term Exclusion Experiment, 11–13, 152–153
Knotted wrack (*Ascophyllum nodosum*), 28–30, 158
Kruger National Park, South Africa, 115–116, 186

L

Lack clutch size, 61–62, 167
Larus glaucescens, *see* glaucous-winged gull
Leaf litter, 40–41, 161
Legumes, 38, 160
Lemon damselfish (*Pomacentrus moluccensis*), 136–138, 193
Lepus americanus, *see* snowshoe hare
Life history, 58–61, 166–167
Lignin, 43–45, 162
Limonium californicum, *see* California sea lavender
Limpet
  checkered (*Lottia strigatella*), 116–121, 186–188
  Lottia spp., 146–148, 196
  ribbed (*Lottia digitalis*), 116–121, 186–188
  shield (*Lottia pelta*), 116–121, 186–188
  spoon (*Scuttelastra cochlear*), 56–58, 166
  warming treatment, 146–148, 196
Linear mixed model, 111–114, 184–185
Linear model, 139–141, 194
Linkage, *see* food web
Lion, 115–116, 185–186
Litter
  bags xvi–xx, 43–45, 161
  processing, 40–41, 43–45, 161, 162
Littoraria irrorata, *see* marsh periwinkle
Lizard, 114–115, 185
Lizard Island, Great Barrier Reef, 23–24, 156
Longevity, 130–132, 190–191
Lotka-Volterra models, 74–75, 172
Lottia digitalis, *see* ribbed limpet
L. pelta, *see* shield limpet
L. strigatella, *see* checkered limpet
Lutra lutra, *see* Eurasian otter
Lynx (*Lynx lynx*), 97–98, 179
Lynx lynx, *see* lynx

M

MacArthur and Wilson equilibrium model, 9, 151
Madeira, 8–9, 151–152
Mallard, 124–127, 189
Mangrove communities, 48–50, 163–164
Mann-Whitney U test, 58–61, 166–167
Manorina melanocephala, *see* noisy miner
Mar Chiquita salt marsh, Argentina, 27–28, 157
Maritrema poulini, *see* trematode flatworm
Marsh dodder (*Cuscuta salina*), 90–91, 177–178
Marsh periwinkle (*Littoraria irrorata*), 81–82, 82–85, 174, 174–175
Masting event, 105–106, 182
Mean xv
Mediterranean oak (*Quercus* spp.) 105–106, 182
Melanoplus femurrubrum, *see* red-legged grasshopper
Metabarcoding, 124–127, 127–129, 189, 189–190

Mite community xxii–xxiii
Modiolus modiolus, *see* horse mussel
Moorea Island, French Polynesia, 25–26, 156–157
Moose (*Alces alces*), 95–97, 179
Moth
  autumn 127–129, 189–190
  box tree 53–56, 165–166
  winter 127–129, 189–190
  yucca 88–89, 176–177
Mountain beech (*Nothofagus cliffortioides*), 7–8, 151
Mountain bluebird (*Sialia currucoides*), 67–69, 169
Mountain Lake, Virginia, US, 87–88, 176
Mpala Research Centre, Laikipia, Kenya, 11–13, 152–153
Muhlenbergia capillaris, *see* pink muhly grass
Multivariate analysis xx–xxiii
Mussel, 16–121, 186–188
Mute swan, 124–127, 189
Mutualism
  ants, 78–80, 173–174
  Branchiobdellida, 87–88, 176
  Breviotum minutum, 85–87, 175–176
  crayfish, 87–88, 176
  Exaiptasia pallida, 85–87, 175–176
  mycorrhiza, 127–129, 134–135, 192–193
  whistling thorn, 78–80, 173–174
  yucca 88–89, 176–177
Mytilus trossulus, *see* blue mussel

N

N
  Hawaiian soils, 37–39, 160–161
  mangroves, 48–50, 163–164
  streams, 33–37, 159–160
  whale faeces, 45–48, 163
^{15}N
  anemones, 85–87, 176
  Arctic food webs, 143–146, 195
  brown bear, 95–97, 179
  mangroves, 48–50, 163–164
  seabirds, 72–73, 171
  whales, 45–48, 163
N:P ratio, 49, 164
Nahant, Massachusetts, US, 28–30, 158
Neogale vison, *see* American mink
New Mexico spadefoot toad (*Spea multiplicata*), 58–61, 166–167
New Zealand mudsnail (*Potamopyrgus antipodarum*), 132–134, 191–192
Niche partitioning, 72, 73, 75–77, 172–173
NMDS, *see* Non-metric multidimensional scaling
Noisy miner (*Manorina melanocephala*), 64–65, 168–169
Non-biting midge, 124–127, 189
Non-metric multidimensional scaling (NMDS)
  African jewelfish 15–19, 153–154
  Interpretation, xx–xxi
  limpet communities, 146–148, 196
  mycorrhizas, 134–135, 192–193
  rocky shore succession, 15–19, 158
Nothofagus cliffortioides, *see* mountain beech
Notophthalmus viridescens, *see* Eastern newt
Nursery web spider (*Pisaurina mira*), 108–110, 183–184
Nutrient cycling, 45–48, 48–50, 163, 163–164

O

O_2 consumption, 39–40, 161
Oak (*Quercus* spp.), 105–106, 182
Omnivory, 58–61, 166–167
Oniscus asellus, *see* woodlouse
Opherophtera brumata, *see* moth, winter

Oystercatcher (*Haematopus ostralegus*), 63–64, 168

P

P
- *Daphnia*, 69–70, 170
- Hawaiian soils, 37–39, 160–161
- mangroves, 48–50, 163–164
- streams, 34–37, 159–160
- whale faeces, 45–48, 163

Pairwise comparison xvii–xviii, 175
Paramecium spp., 74–75, 172
Parasitism
- avian vampire fly, 91–94, 178
- crayfish, 87–88, 176
- mistletoe, 192–193
- mudsnails, 132–134
- sea anemones, 85–87, 175–176
- trematodes, 132–134
- yucca 88–89, 176–177

Parorchis acanthus, see trematode worm
Pathway model, 111–114, 184–185
PCA, *see* Principal components analysis
Phenolics, 43–45,127–129, 162, 189–190
Philornis downsii, see avian vampire fly
Pickleweed (*Salicornia perennis*), 27–28, 157
Pickleweed (*Salicornia virginica*), 90–91, 177–178
Pink muhly grass (*Muhlenbergia capillaris)*, 66–67,169
Pinus edulis, see pinyon pine
Pinyon pine (*Pinus edulis*), 132–134, 192–193
Pisaurina mira, see nursery web spider
Plant functional groups, 1–4, 150
Poa pratensis, see smooth meadow grass
Pollicipes spp., *see* barnacle, goose
Pomacentrus amboinensis, see Ambon damselfish
Pomacentrus moluccensis, see lemon damselfish
Population cycles
- Arctic ground squirrel 97–98, 179
- *Capitella capitata*, 51–52, 164
- lynx, 97–98, 179
- snowshoe hare, 97–98, 179

Population density, 139–141, 194
Population growth
- exponential, 52–53, 164
- logistic 52–53, 164

Population time lag, 51–53, 165
Potamopyrgus antipodarum, see New Zealand mudsnail
Predation
- African jewelfish, 15–19, 153
- Arctic, 143–146, 195
- brown bears, 95–97, 179
- coastal communities, 116–121, 186–188
- culling, 139–141, 167, 168–169, 194
- eider, 139–141, 194
- grasslands, 108–110, 183–184
- Kruger National Park, 115–116, 185–186
- moose, 95–97, 179
- predator-prey cycle, 103–104, 181
- predator satiation, 179, 182
- prey switching, 97–98, 179
- seeds, 9–11, 152
- stream communities, 34–37, 159–160

Predator-prey cycles, 103–104, 181–182
Preference-performance hypothesis, 55, 165–166
Primary consumer, 4–6, 159–160
Primary producer, 4–6, 159–160
Primary productivity, 108–110, 183–184
Principal component analysis (PCA), xxii–xxiii, 45–48, 163
Priority effects, 154, 170
Processing, litter, 40–41, 43–45, 161, 162
Proteas, 19–20, 155

Prunus serotina, *see* black cherry
Pycnopsyche spp., *see* caddis fly
Python bivittatus, *see* Burmese python
P. molurus, *see* Burmese python

Q

Quercus spp., *see* Mediterranean oak

R

Razorbill, 72–73, 171
Red-footed booby, 48–50, 163–164
Red-legged grasshopper (*Melanoplus femurrubrum*)108–110, 183–184
Reproductive rate, 130–132, 190–191
Resilience, 1–4, 150
Resorption efficiency, 48–50, 163–164
Resource partitioning, *see* Niche partitioning
Rhino, 115–116, 185–186
Ringed seal, 143–146, 195
Roan, 115–116, 185–186
Rotifer, 103–104, 181–182
Rubricatochromis letourneuxi, *see* African jewel fish

S

Salamander, 33–37, 159–160
Salicornia perennis, *see* pickleweed
Salicornia virginica, *see* pickleweed
Salt marsh
  herbivory, 81–82, 82–85, 174, 174–175
  marsh dodder, 177–178, 90–91
  oystercatcher, 63–64, 168
  parasitism, 82–85, 174–175
  succession, 27–28, 157
Salt marsh cordgrass (*Sporobolus alterniflora*), 81–82, 82–85, 174, 174–175
Salt Spring Island, British Columbia, Canada, 146–148, 196
Sarcocornis perennis, *see* pickleweed
Scale insects, 78–80, 173–174
Scavenger, 122–124, 188–189
Scutellastra cochlear, *see* limpet, spoon
Sea anemone (*Exaiptasia pallida*), 85–87, 175–176
Sea grape (*Coccoloba uvifera*), 114–115, 185
Seashore salt grass (*Distichlis spicata*), 26, 28, 157
Selection-abortion hypothesis, 88–89, 177
Selective advantage, 116–121, 186–188
Semibalanus balanoides, *see* barnacle, common
Shannon diversity index, 127–129, 189–190
Sialia currucoides, *see* mountain bluebird
S. mexicana, *see* western bluebird
Site fidelity, 130–132, 190–191
Small Indian mongoose (*Urva auropunctata*), 98–99, 189
Smooth meadow grass (*Poa pratensis*), 108–110, 183–184
Snowshoe hare (*Lepus americanus*), 97–98, 179
Soil weathering, 37–39, 160–161
Solidago rugosa, *see* goldenrod
Somateria mollissima, *see* eider
Spartina alterniflora, *see* salt marsh cordgrass
S. densiflora, *see* denseflower cord grass
Spea multiplicata, *see* New Mexico spadefoot toad
Specialist, 130–132, 190–191
Spider, 114–115, 185
Spiny stream crayfish (*Faxonius cristavarius*), 87–88, 176
Sporobolus alterniflora, *see* salt marsh cordgrass
S. montevidensis, *see* denseflower cord grass
Spotted hyaena, 115–116, 185–186

Spotted knapweed (*Centaurea stoebe*), 6–7, 151
Stability of communities, 1–4, 150
Stable coexistence, 75–77, 172–173
Stable isotope analysis
  Arctic food webs, 143–146, 195
  bear diet, 95–97, 179
  seabirds, 72–73, 171
  mangroves, 48–50, 163–164
Stacked graphs, 19–20
Standard deviation (SD), xv–xx, 163–164
Standard error (SE), xv–xx
Staniel Cay, Bahamas, 114–115, 185
Strangford Lough, Northern Ireland, UK, 99–101, 180
Succession
  Connell and Slatyer succession model, 28, 30
  facilitation, 28, 30, 157, 158
  inhibition, 28, 30, 157, 158
  rocky shore, 28–30, 158
  salt marsh, 27–28, 157
  tolerance, 28, 30, 158
Svalbard, 139–141, 194
Sweet wattle (*Acacia suaveolens*), 21–23, 155–156

T

Tegeticula yuccasella, *see* Yucca moth
Termites, 43–45, 162
Thick-billed murre, 72–73, 171
Top-down processes, 72–73, 113, 171, 184
Trace elements, 45–48, 188–189
Trade-off
  bird clutch size, 61–62, 167
  competition-colonisation, 67–69, 69–70, 169, 170, 171
  corals, 24, 156
  life histories, 58–61, 166
Trematode flatworm
  Apatemon sp., 132–134, 191–192
  Aporocotylid sp., 132–134, 191–192
  Coitocaecum parvum, 132–134, 191–192
  Maritrema poulini, 132–134, 191–192
  Parorchis acanthus, 82–85, 174–175
Trophic cascade, 99–101, 180
Trophic decoupling, 36, 159–160
Tsuga diversifolia, *see* Japanese hemlock
Tukey post hoc test
  anemone symbiosis, 85–87, 175–176
  golden rod, 108–110, 183–184
  litter bag experiment xviii–xx
  marsh periwinkle, 82–85, 174–175

U

Urocitellus parryii, *see* Arctic ground squirrel
Ursus arctos, *see* brown bear
Urva auropunctata, *see* small Indian mongoose

V

Vacant niche, 153, 190–191
Vaccinium myrtillus, *see* bilberry
Vachellia drepanolobium, *see* whistling thorn

W

Warming effects on tropical soil, 141–143, 194–195
Warming treatment, 146–148, 196
Western bluebird (*Sialia mexicana*), 67–69, 169
Whales, 45–48, 163
Whistling thorn (*Vachellia drepanolobium*), 78–80, 173–174
Wild boar, 122–124, 188–189
Wild dog, 115–116, 185–186

Wood pigeon (*Columba palumbus*), 62–63, 124–127, 167–168, 189
Woodlouse (*Oniscus asellus*), 40–41, 161

Y

Yucca (*Yucca filamentosa*), 88–89, 176–177
Yucca filamentosa, *see* Yucca

For Product Safety Concerns and Information please contact our EU representative GPSR@taylorandfrancis.com Taylor & Francis Verlag GmbH, Kaufingerstraße 24, 80331 München, Germany